Springer Theses

Recognizing Outstanding Ph.D. Research

For further volumes:
http://www.springer.com/series/8790

Aims and Scope

The series "Springer Theses" brings together a selection of the very best Ph.D. theses from around the world and across the physical sciences. Nominated and endorsed by two recognized specialists, each published volume has been selected for its scientific excellence and the high impact of its contents for the pertinent field of research. For greater accessibility to non-specialists, the published versions include an extended introduction, as well as a foreword by the student's supervisor explaining the special relevance of the work for the field. As a whole, the series will provide a valuable resource both for newcomers to the research fields described, and for other scientists seeking detailed background information on special questions. Finally, it provides an accredited documentation of the valuable contributions made by today's younger generation of scientists.

Theses are accepted into the series by invited nomination only and must fulfill all of the following criteria

- They must be written in good English.
- The topic should fall within the confines of Chemistry, Physics, Earth Sciences, Engineering and related interdisciplinary fields such as Materials, Nanoscience, Chemical Engineering, Complex Systems and Biophysics.
- The work reported in the thesis must represent a significant scientific advance.
- If the thesis includes previously published material, permission to reproduce this must be gained from the respective copyright holder.
- They must have been examined and passed during the 12 months prior to nomination.
- Each thesis should include a foreword by the supervisor outlining the significance of its content.
- The theses should have a clearly defined structure including an introduction accessible to scientists not expert in that particular field.

Yan Zeng

Colloidal Dispersions Under Slit-Pore Confinement

Doctoral Thesis accepted by
Technical University of Berlin, Germany

 Springer

Author
Dr. Yan Zeng
North Carolina State University
Raleigh, NC
USA

Supervisor
Prof. Regine von Klitzing
Fakultät II/Institut für Chemie
Technische Universität Berlin
Berlin
Germany

ISSN 2190-5053   ISSN 2190-5061   (electronic)
ISBN 978-3-642-43712-0   ISBN 978-3-642-34991-1 (eBook)
DOI 10.1007/978-3-642-34991-1
Springer Heidelberg New York Dordrecht London

Printed on acid-free paper

Springer is part of Springer Science+Business Media (www.springer.com)

Related Publications

Surviving Structure in Colloidal Suspensions Squeezed from 3D to 2D, Sabine H. L. Klapp, Yan Zeng, Dan Qu, and Regine von Klitzing, *Physical Review Letters*, **2008**, *100*, 118303

Asymptotic structure of charged colloids between two and three dimensions: the influence of salt, Sabine H. L. Klapp, Stefan Grandner, Yan Zeng, and Regine von Klitzing, *Journal of Physics: Condensed Matter*, **2008**, *20*, 494232

Impact of surface charges on the solvation forces in confined colloidal solutions, Stefan Grandner, Yan Zeng, Regine von Klitzing, and Sabine H. L. Klapp, *The Journal of Chemical Physics*, **2009**, *131*, 154702

Oscillatory Structural Forces Due to Nonionic Surfactant Micelles: Data by Colloidal-Probe AFM vs Theory, Nikolay C. Christov, Krassimir D. Danov, Yan Zeng, Peter A. Kralchevsky, and Regine von Klitzing, *Langmuir*, **2010**, *26*, 915–923

Charged silica suspensions as model materials for liquids in confined geometries, Sabine H. L. Klapp, Stefan Grandner, Yan Zeng, and Regine von Klitzing, *Soft Matter*, **2010**, *6*, 2330–2336

Effect of particle size and Debye length on order parameters of colloidal silica suspensions under confinement, Yan Zeng, Stefan Grandner, Cristiano L.P. Oliveira, Andreas F. Thuenemann, Oskar Paris, Jan S. Pedersen, Sabine H. L. Klapp, and Regine von Klitzing, *Soft Matter*, **2011**, *7*, 10899–10909

Structuring of colloidal suspensions confined between a silica microsphere and an air bubble, Yan Zeng, Regine von Klitzing, *Soft Matter*, **2011**, *7*, 5329–5338

Oscillatory Forces of Nanoparticle Suspensions Confined between Rough Surfaces Modified with Polyelectrolytes via the Layer-by-Layer Technique, Yan Zeng, Regine von Klitzing, *Langmuir*, **2012**, *28*, 6313–6321

To my family

Supervisor's Foreword

Interactions between colloidal particles address a fundamental topic related to the aim of stabilizing macroscopic colloidal dispersions like foams, emulsions, and suspensions. Due to ongoing miniaturization in many technologies, the structuring of colloidal particles under confinement is attracting more and more attention in research nowadays.

In her thesis, Yan Zeng produced the confinement in a Colloidal Probe AFM between a silica microsphere and a planar silicon wafer presenting a slit-pore geometry. Measuring the force through complex fluids like silica suspensions and micellar solutions leads to oscillatory force curves, also called as structural forces due to a layerwise expulsion of the particles. The resulting characteristic of two parameters (particle distance and correlations length) are in very good agreement with the ones obtained from respective scattering data (SAXS) of the bulk phase. There is a fundamental difference between the distance of neutral particles (micelles of nonionic surfactants) and charged particles (negatively charged silica particles)—Neutral particles can be pressed in contact under confinement, while charged particles are pressed out of the slitpore before coming into contact. The distance between the charged nanoparticles scales with the concentration c with $c^{-1/3}$ and not with (diameter of particle + 2 × Debye length) as often mentioned in literature. The scaling behavior is very robust against changes in parameters like ionic strength and particle size. The correlation length scales with (radius of particle + Debye length).

One focus of the thesis was the separation between the confinement effect itself and the effect of the properties of the confining surfaces. Therefore, the silicon wafer was coated with polyelectrolytes or it was replaced by the fluid interface of an air bubble. Both the particle distance and the correlation length remain constant, and only the interaction strength is affected by the surface roughness, surface potential, and/or surface elasticity.

This thesis is an excellent example for a successful collaboration between theoreticians and experimentalists. The group of Sabine Klapp at TU Berlin showed that the experimental data can be reproduced by Monte Carlo simulations and with a DLVO-like model. Yan Zeng showed that the wavelength of the force oscillation for the asymptotic distance regime is identical with one of the pair correlation function of the volume phase. This was predicted by density functional theory long ago, but had not been confirmed before experimentally. In the thesis of Yan Zeng this fundamental prediction is verified for the first time for a real colloidal system.

Berlin, September 2012   Regine von Klitzing

Acknowledgments

The presented thesis is based on the experiments performed during my time as a Ph.D. student at the Stranski Laboratory of Physical and Theoretical Chemistry at TU Berlin.

I want to thank my advisor Prof. Regine von Klitzing for offering me the Ph.D. position and the interesting research topic. She has always welcomed discussion of all kinds of experimental related questions and constantly brought me many helpful suggestions. She encouraged me through her own personal strength and gave me a lot of opportunities to present my work at conferences. Besides, she was warm and helpful on my personal life.

I also thank my old boss Prof. Findenegg for recommending me to Prof. Regine von Klitzing after I finished my Masters study, and I thank his constant interest in my research and his care for my personal life.

Professor Sabine Klapp from TU Berlin is much acknowledged for her discussion on the theoretical simulations and her effort on helping me to understand better the basis of the research background. One part of the presented thesis arose from the fruitful cooperation with her group. Her Ph.D. student Stefan Grandner is also acknowledged for his helpful discussion and the friendly acquisition of accompanying papers.

Professor Otto Glatter from University of Graz, Prof. Jan S. Pederson from University of Aarhus, Prof. Oscar Paris from Montanuniversitaet Leoben (worked formerly in MPI, Golm), and Dr. Andreas F. Thünemann from BAM Berlin are kindly acknowledged for the implementation of X-ray scattering experiments and for their valuable support during the evaluation.

I also acknowledge Prof. Mark Rutland from Royal Institute of Technology for being my second supervisor. I want to thank Prof. Dr. Reinhard Schomäcker from TU Berlin for agreeing to be the "Vorsitzenderin" in my defense.

I would like to thank all employees of the Stranski laboratory for the harmonious working environment and numerous large and small aids. I especially want to acknowledge the "AFM" force measurement members: Cagri Üzüm, Liset Lüderitz, and Sebastian Schön for fruitful discussions and suggestions, and moreover, for their friendship. Dr. Sylvain Prévost, Bhuvnesh Bharti, and Stefan

Wellert for the advice and help on using of the SAS-fit software. The technicians Gabi Hedicke, Michaela Dzionara, and Ingke Ketelsen for their help in my laboratory work. Dr. Rene Straßnick for solving the computer and software problems. Nora Kristen and Natascha Schelero for creating nice laboratory environment in the same measuring room. My office members Samuel Dodoo, Cagri Üzüm, and Johannes Hellwig for encouraging on the work and writing as well as by bringing a lot of pleasures on daily life.

In addition, Dersy Lugo, Dilek Akcakayiran, Michael Muthig, Rastko Joksimovic, Hsin-yi Liu, Mengjun Xue, and all those friends in the Stranski laboratory are acknowledged for their friendship and support.

My Mum, Dad, and my brother Chao for motivating me and being always there supporting me. Christopher Popeney, for his invaluable mental support, understanding, and proofing.

The financial support from the Deutsche Forschungsgemeinschaft under the frame of the priority program Sfb 448 and SPP 1273 "Kolloidverfahrenstechnik" (KL1165/10-1) as well as TU Berlin through Frauen-Promotionsabschlussstipendium are kindly acknowledged.

Thank you.

Contents

Abbreviations and Symbols[1]

A	Amplitude of the oscillatory structural force
c	Weight percentage of silica nanoparticles
c_s	Molar concentration of surfactant
CMC	Critical micelle concentration
d	Thickness of layers in a layer-by-layer coated polymeric film
d	Micelle diameter
F	Force
F/R	Normalized force measured by AFM
F_{Ta}	Hydrodynamic force
$g(r)$	Bulk pair correlation function
h	Separation between AFM colloidal probe and the substrate
I_{max}	Maximum scattering intensity in arbitrary units
I_{tot}	Total ionic strength
I_{salt}	Ionic strength of added salt
k_b	Stiffness of deformable surface
k_c	AFM cantilever spring constant
K	Conductivity
κ^{-1}	Debye length
κ_W^{-1}	Debye length including the wall-counterions
N	Number of layers in a layer-by-layer coated polymeric film
$P_{\max}$	Height of the first maximum of the solvation pressure
q_{max}	Momentum transfer in the position of maximum scattering
q^{-1}	Dimensionless correlation length of micelle
Δq	Full width at half maximum of the structure peak
R	Radius of the AFM colloidal probe
R	Radius of the silica nanoparticle
$2(R + \kappa^{-1})$	Effective particle diameter
R_{RMS}	Root mean square roughness

---

[1] Non-essential and empirical parameters are not included.

t^{-1}	Decay length of electrostatic repulsive force
u	Scanning velocity
w_0	Dimensionless interaction strength of micelle
ΔX	Change in the nominal separation
Z	Valency of charged particle
$\tilde{Z}$	Effective valency of charged particle
Z_c	AFM cantilever deflection
ζ	Zeta-potential
ψ_S	Surface potential of confining wall
ϕ	Volume fraction
δ	Deformation
θ_f	Phase shift in the oscillatory force (solvation force) curve
θ	Contact angle
η	Viscosity
λ	Wavelength
λ_f	Wavelength in confinement
λ_b	Wavelength in bulk
$2\pi/\omega$	Dimensionless wavelength of micelle
ρ	Particle number density
ρ_c	Number density of counterions
ρ_p	Mass density of silica particle
ρ_s	Mass density of solution
σ	Diameter of the silica nanoparticle
γ	Surface tension
τ	Relaxation time
ξ	Correlation length
ξ_f	Correlation length in confinement
ξ_b	Correlation length in bulk
AFM	Atomic force microscope/microscopy
Brij 35	Polyoxyethylene lauryl ether
$C_{16}TAB$	Hexadecyltrimethylammonium bromide
$\beta\text{-}C_{12}G_2$	β-dodecylmaltoside
CP-AFM	Colloidal probe atomic force microscope/microscopy
DLVO	Derjaguin-Landau-Verwey-Overbeek
GCMC	Grand canonical Monte Carlo
HA	Hyaluronic acid
HNC	Hypernetted chain
HS 40	Silica particle suspensions with diameter of 16 nm
InvOLS	Deflection inverse optical lever sensitivity
MC	Monte Carlo
PAA	Poly(acrylic acid)
PAH	Poly(allylamine hydrochloride)
PEI	Polyethylenimin
PSS	Poly(sodium 4-styrenesulfonate)

SAXS	Small angle X-ray scattering
SDS	Sodium dodecyl sulfate
SM 30	Silica particle suspensions with diameter of 11 nm
Tween 20	Polyoxyethylene sorbitan monolaurate
TFPB	Thin film pressure balance
TMA 34	Silica particle suspensions with diameter of 26 nm

Chapter 1
Introduction

Colloidal suspensions are solutions containing small particles. These particles are larger than the molecules of the medium. They have a typical size ranging from several 10 nanometers to micrometers. They can be made from different materials and suspended in a wide variety of solvents. Colloidal dispersions have large application in our daily life, for instance, as cosmetics, advanced ceramics [1, 2], coating [3], paints, and inks [4, 5]. Also, thin films of colloidal dispersions are confined to a solid substrate to manufacture advanced self-assembled materials such as photonic crystals [6–10], and sensors [11, 12]. In addition, special colloids have biological applications, e.g. as pharmaceuticals, in drug delivery [13], and in food processing.

A distinguishing feature of colloidal systems is that the contact area between dispersed phase and the dispersing medium is large. As a result, surface forces strongly influence dispersion behavior. By tailoring interactions between dispersed-phase particles, one can design colloids needed for specific applications. The stability, rheology, and other desired properties of the colloids are externally controlled by the surface charge of colloids and the properties of the dispersing medium, such as temperature, pH, and ionic strength. The attractive van der Waals forces are ubiquitous in colloidal systems and must be balanced by electrostatic, steric, or other repulsive interactions to achieve the desired degree of colloidal stability.

Many applications of colloids also involve interactions on substrates, for example, the spreading and adhesion phenomena on the substrates (e.g. paints and inks) or the transport of colloids in micro fluidic devices or in porous media or at biological surfaces (e.g. drug delivery, stacking of red blood cells). Besides, an enhanced interest in miniaturization of the producing process draws increasing research attention. Therefore the confinement effects on the colloidal structuring formation and properties are needed to be considered.

The confinement induces colloids to order differently as compared to the bulk. Colloidal crystal structure has been reported [14, 15] in confinement at sufficiently high concentration. At relatively low concentrations, layered ordering of colloids has been found [14, 16–31]. The first study of the ordering of colloidal particles can be traced back to the 1980s. Nikolov found that thinning films of aqueous dispersions

Y. Zeng, *Colloidal Dispersions Under Slit-Pore Confinement*, Springer Theses,
DOI: 10.1007/978-3-642-34991-1_1, © Springer-Verlag Berlin Heidelberg 2012

of polystyrene latex nanoparticles changed thickness with regular step-wise abrupt transitions by using reflected light microinterferometry [14]. These observations verified that the step-wise thinning or stratification of thin liquid films could be explained as a layer-by-layer thinning of ordered structuring of colloidal particles formed inside the film. There are several other papers that have also shown that particles tend to form periodic ordering during the approach of confining surfaces by methods of thin film pressure balance [16–18] and total reflectometry [19, 20].

Recently, the structuring formation has been studied by the measurement of the oscillatory force of colloidal particles by Piech and Drelich et al. [21–25] with colloidal probe atomic force microscopy (CP-AFM), which was developed by Ducker and Butt [32, 33] and was advantageous in measuring the complete oscillatory force curves for various systems [21–25, 30, 34–38]. The concentration profile of the particles oscillates with the separation from one confining surface. The oscillatory force occurs when the oscillating concentration profile of the particles in front of the opposing confining surfaces overlap. With decreasing separation between the two confining surfaces, the layers of particles are pressed out one after another, which leads to measurable alternating repulsion and attraction. The oscillatory force thus indicates the periodic layering of confined colloids. The oscillatory wavelength represents the distance between two adjacent layers of particles formed parallel to the confining surfaces. The decay length is a measure of how far particles correlate to obtain periodic oscillations.

Previous work has focused on revealing the dependency of inter-colloid distance in confinement on the colloid concentration. However there are still numerous questions left to be answered. How does the confinement effect behave on the two characteristic lengths: the inter-colloid distance and the correlation length? What is the dependency of characteristic lengths on the colloid size? Is the structuring of colloids in confinement affected by the total ionic strength of the dispersions? What is the effect of confining surface charge/potential on the corresponding structuring? Can oscillatory force still be observed on rough or deformable confining surfaces? How does the surface charge and deformability of colloids influence the corresponding structuring?

This thesis is aimed to answer the above questions. The structuring of colloids under confinement is studied via force measurements with CP-AFM. In the following chapter, a brief theoretical overview on the different interactions present in the investigated systems are discussed and the computational simulation for the investigated systems are described. Then in Chap. 3, the basic principles of the used experimental technique are given, with emphasis on its advantages, which make it the technique of choice for the investigation. In Chap. 4, the structuring of silica nanoparticles confined between two smooth, solid surfaces is investigated, subdivided into three parts based on the effect of particle concentration, ionic strength, and particle size. The characteristic quantities, i.e. interparticle distance, correlation length, and interaction strength are extracted from the oscillatory force and their relation with the system parameters is investigated. Additionally, SAXS measurements, Monte Carlo simulations and hypernetted chain (HNC) calculations are included in order to examine in detail the effect of confinement on the characteristic quantities. Chapter 5 discusses the influence of surface potential and roughness of the confining surfaces on the structuring of nanoparticles. Monte Carlo simulation with a modified particle-wall potential are combined to reveal the mechanism behind the change in force amplitude. In Chap. 6, the effect of the surface deformability on the structuring of silica nanoparticles is investigated by confining nanoparticles between a solid sphere and an air bubble. Various surfactants are used to tune the bubble deformability. Chapter 7 describes the influence of the surface charge and deformability of colloids on their structuring formation by probing non ionic surfactant micelles. Theoretical calculations based on the hard-sphere model is combined to provide a deeper understanding of the observed processes.

References

1. Lewis, J. (2000). *Journal of the American Ceramic Society, 83*, 2341–2359.
2. Tohver, V., Smay, J., Braem, A., Braun, P., & Lewis, J. (2001). *Proceedings of the National Academy of Sciences of the United States of America, 98*, 8950–8954.
3. Agarwal, N., & Farris, R. (2000). *Polymer Engineering and Science, 40*, 376–390.
4. Chrisey, D. (2000). *Science, 289*, 879.
5. Smay, J., Cesarano, J., & Lewis, J. (2002). *Langmuir, 18*, 5429–5437.
6. Yablonovitcah, E. (1987). *Physical Review Letters, 58*, 2059–2062.
7. Joannopoulos, J., Villeneuve, P., & Fan, S. (1997). *Nature, 386*, 143–149.
8. Pan, G., Kesavamoorthy, R., & Asher, S. (1997). *Physical Review Letters, 78*, 3860–3863.
9. Braun, P., & Wiltzius, P. (1999). *Nature, 402*, 603–604.
10. Johnson, S., Ollivier, P., & Mallouk, T. (1999). *Science, 283*, 963–965.
11. Tressler, J., Alkoy, S., Dogan, A., & Newnham, R. (1999). *Composites Part A, 30*, 477–482.
12. Allahverdi, M., Danforth, S., Jafari, M., & Safari, A. (2001). *Journal of the European Ceramic Society, 21*, 1485–1490.
13. Garnett, M. C., Stolnick, S., Dunn, S. E., Armstrong, I., Ling, W., Schacht, E., et al. (1999). *Materials Research Society Bulletin, 24*, 49–56.
14. Nikolov, A., & Wasan, D. (1989). *Journal of Colloid and Interface Science, 133*, 1–12.
15. Wasan, D., & Nikolov, A. (2003). *Nature, 423*, 156–159.
16. Basheva, E., Danov, K., & Kralchevsky, P. (1997). *Langmuir, 13*, 4342–4348.
17. Sethumadhavan, G., Nikolov, A., & Wasan, D. (2001). *Journal of Colloid and Interface Science, 240*, 105–112.
18. Denkov, N., Yoshimura, H., Nagayama, K., & Kouyama, T. (1996). *Physical Review Letters, 76*, 2354–2357.
19. Sharma, A., & Walz, J. (1996). *Journal of the Chemical Society, Faraday Transactions, 92*, 4997–5004.
20. Sharma, A., Tan, S., & Walz, J. (1997). *Journal of Colloid and Interface Science, 191*, 236–246.
21. Piech, M., & Walz, J. (2002). *Journal of Colloid and Interface Science, 253*, 117–129.
22. Piech, M., & Walz, J. (2004). *Journal of Physical Chemistry B, 108*, 9177–9188.
23. McNamee, C., Tsujii, Y., Ohshima, H., & Matsumoto, M. (2004). *Langmuir, 20*, 1953–1962.
24. Tulpar, A., Van Tassel, P., & Walz, J. (2006). *Langmuir, 22*, 2876–2883.
25. Drelich, J., Long, J., Xu, Z., Masliyah, J., Nalaskowski, J., Beauchamp, R., et al. (2006). *Journal of Colloid and Interface Science, 301*, 511–522.
26. Bergeron, V., & Radke, C. (1992). *Langmuir, 8*, 3020–3026.
27. Bergeron, V., Jimenezlaguna, A., & Radke, C. (1992). *Langmuir, 8*, 3027–3032.
28. Richetti, P., & Kekicheff, P. (1992). *Physical Review Letters, 68*, 1951–1954.

29. Parker, J., Richetti, P., Kekicheff, P., & Sarman, S. (1992). *Physical Review Letters, 68,* 1955–1958.
30. McNamee, C., Tsujii, Y., & Matsumoto, M. (2004). *Langmuir, 20,* 1791–1798.
31. Nikolov, A., Kralchevsky, P., Ivanov, I., & Wasan, D. (1989). *Journal of Colloid and Interface Science, 133,* 13–22.
32. Ducker, W., Senden, T., & Pashley, R. (1991). *Nature, 353,* 239–241.
33. Butt, H. (1991). *Biophysical Journal, 60,* 1438–1444.
34. Milling, A. (1996). *Journal of Physical Chemistry, 100,* 8986–8993.
35. Biggs, S., Burns, J., Yan, Y., Jameson, G., & Jenkins, P. (2000). *Langmuir, 16,* 9242–9248.
36. Biggs, S., Prieve, D., & Dagastine, R. (2005). *Langmuir, 21,* 5421–5428.
37. Qu, D., Baigl, D., Williams, C., Mohwald, H., & Fery, A. (2003). *Macromolecules, 36,* 6878–6883.
38. Qu, D., Pedersen, J. S., Garnier, S., Laschewsky, A., Moehwald, H., & von Klitzing, R. (2006). *Macromolecules, 39,* 7364–7371.

Chapter 2
Scientific Background

A colloidal system consists of two separate phases: a dispersed phase (or internal phase) and a continuous phase (or dispersion medium). The dispersed phase and the continuous medium can be in gas, liquid, and solid states. The dispersed-phase has a diameter of between approximately 1 and 1000 nm. Homogeneous mixtures with a dispersed phase in this size range may be called colloidal aerosols, colloidal emulsions, colloidal foams, colloidal suspensions, or hydrosols depending on varying combinations of dispersed phase and continuous phase.

A distinguishing feature of colloidal systems is that the contact area between dispersed phase and the dispersing medium is large. As a result, surface forces strongly influence dispersion behavior. By tailoring interactions between dispersed-phase, one can design colloids needed for specific applications. The stability, rheology, and other desired properties of colloids are controlled internally by the surface charge of the dispersed-phase and externally by the properties of the dispersing medium, such as temperature, pH, and ionic strength.

2.1 Surface Forces

The forces of charged colloids interacting through a liquid medium can be described by Derjaguin-Landau-Verwey-Overbeek (DLVO) theory [1, 2]. It combines the effects of the van der Waals attraction and the electrostatic repulsion due to the so-called double layer of counterions. Because of the markedly different distance dependency of the van der Waals and electrostatic interactions, the total force can show several minima and maxima with varying interparticle distance. Additional forces, such as structural force (solvation force) and hydrophobic force commonly occur in aqueous solutions. Considering the present experimental systems, the involved forces will be described: van der Waal force, electrostatic force, structural force and hydrophobic force.

Y. Zeng, *Colloidal Dispersions Under Slit-Pore Confinement*, Springer Theses,
DOI: 10.1007/978-3-642-34991-1_2, © Springer-Verlag Berlin Heidelberg 2012

2.1.1 Van der Waals Force

Van der Waals forces are a family of short-range forces, including the dipole–dipole force, dipole-induced dipole force, and dispersion forces. The expression of the van der Waals interaction between particles can follow the method by Hamaker [3], in which the net interaction energy is the integration of all pair contributions between two bodies. Thus the non-retarded van der Waals interaction energy of two spheres of radius R_1 and R_2 can be obtained as

$$V_{vdW}(D) = -\frac{A_H}{6}\left[\frac{2R_1R_2}{D^2 + 2(R_1 + R_2)D} + \frac{2R_1R_2}{D^2 + 2(R_1 + R_2)D + 4R_1R_2}\right.$$
$$\left. + \ln\left(\frac{D^2 + 2R_1D + 2R_2D}{D^2 + 2(R_1 + R_2)D + 4R_1R_2}\right)\right] \tag{2.1}$$

where A_H is the Hamaker constant, R_1 and R_2 are the radius of particle 1 and 2, respectively, and D is the surface-to-surface distance, that is, $D = r - (R_1 + R_2)$ (r being the particle center-to-center distance).

The corresponding simplified expression of the van der Waals interaction energy of a particle approaching a surface is

$$V_{vdW}(D) = -\frac{A_H R_{norm}}{6D} \tag{2.2}$$

where R_{norm} is the normalized radius, which depends on the geometry used. In the case of sphere/flat geometry: $R_{norm} = R_{sphere}$, sphere/sphere geometry: $R_{norm} = R_1R_2/(R_1 + R_2)$. The van der Waals force can be obtained by differentiating the energy with respect to distance

$$F_{vdW} = -\frac{dV_{vdW}}{dD} = \frac{A_H R_{norm}}{6D^2} \tag{2.3}$$

The van der Waals force is always attractive between identical surfaces of the same materials, and can be repulsive between surfaces of dissimilar materials. Hamaker's method and the associated Hamaker constant A_H assumes that the interaction is pairwise additive and ignores the influence of an intervening medium between the two particles of interaction.

$$A_H = \pi^2 \times C \times \rho_1 \times \rho_2 \tag{2.4}$$

where ρ_1 and ρ_2 are the number of atoms per unit volume in two interacting bodies and C is the coefficient in the particle-particle pair interaction. The more advanced Lifshitz theory [4] has a same expression of the van der Waals energy but with consideration of the dielectric properties of the intervening medium, thus the Hamaker constant has a different value.

2.1.2 Electrostatic Force

The electrostatic force originates from the fact that most surfaces in contact with any liquid of high dielectric constant acquire a surface charge. The surface can either be charged by ionization of surface groups (e.g. silanol groups for glass or silica surfaces) or by adsorption of charged ions from the surrounding solution. This results in the development of a surface potential which attracts counterions from the surrounding solution and repels co-ions. In equilibrium, the surface charge is electrically neutralized by oppositely charged counterions in solution within some distance from the surface. The region near the surface of enhanced counterion concentration is called the electrical double layer (EDL) [1]. The EDL can be approximately sub-divided into two regions. Ions in the region closest to the charged wall surface are strongly bound to the surface. This immobile layer is called the Stern or Helmholtz layer. The region adjacent to the Stern layer is called the diffuse layer and contains loosely associated ions that are comparatively mobile. The whole electrical double layer, due to the distribution of the counterion concentration, results in the electrostatic screening of the surface charge.

The decay length of the diffuse electric double layer is known as the Debye screening length [5], κ^{-1}, which is purely a property of the electrolyte solution. The Debye length falls with increasing ionic strength of the solution. In totally pure water at pH 7, κ^{-1} is 960 nm, and in 1 mM NaCl solution κ^{-1} is 9.6 nm. The value κ is given by the relation

$$\kappa = \sqrt{\sum_i \rho_{\infty i} e_0^2 z_i^2 / \epsilon \epsilon_0 k_B T} \tag{2.5}$$

where $\rho_{\infty i}$ is the number density of ion i in the bulk solution, z_i is the valency of the ion i, e_0 is the elementary charge, ϵ_0 and ϵ are the permittivity of the vacuum and the solvent dielectric constant, respectively, and k_B is Boltzmann's constant.

The Debye length determines the range of the electrostatic double-layer interaction between two charged surfaces. The repulsive interaction between two equally charged surfaces is an entropic (osmotic) force. Actually, the electrostatic contribution to the net force is attractive. What maintains the diffuse double layer is the repulsive osmotic pressure between the counterions which forces them away from the surfaces and from each other so as to increase their configurational entropy. When bringing two equally charged surfaces together, one is therefore forcing the counterions back onto the surfaces against their osmotic repulsion, but favored by the electrostatic interaction. The former dominates and the net force is repulsive. Commonly speaking, when two charge surfaces approach each other, the electric double layers overlap and results in the so-called electric or electrostatic double layer repulsion force, even though the repulsion really arises from entropic confinement of the double layer ions.

The electrostatic interaction can be obtained by solving the Poisson-Boltzmann equation [6] for the potential distribution or counterion distribution in the liquid, subject to suitable boundary conditions [5, 7]. These conditions are usually either constant surface potential if the concentration of counterions is constant as D is decreased (e.g. metal sols in a solution) or constant surface charge if the total number of counterions in liquid does not change (e.g. clay minerals). Using weak overlap approximation at constant potential [5], the free energy per unit area of interaction between two spheres is

$$W_{EL} = (64 k_B T \rho_\infty \gamma_1 \gamma_2 / \kappa) e^{-\kappa D} \tag{2.6}$$

where γ is the reduced surface potential $\gamma = \tanh\left(\frac{ze\psi_0}{4k_B T}\right)$, ψ_0 is the potential on the surface.

Using Derjaguin approximation $F = 2\pi R_{norm} W$ [8], the expression of electrostatic force F between two spheres becomes

$$F_{EL} = (128\pi k_B T \rho_\infty R_{norm} \gamma_1 \gamma_2 / \kappa) e^{-\kappa D} \tag{2.7}$$

In the simplest case, $F_{EL} = (64\pi k_B T \rho_\infty R \gamma^2 / \kappa) e^{-\kappa D}$ for two identically charged particles of radius R. This approximation is appropriate for surface potential between 30 and 100 mV. At low surface potentials, below 30 mV, the electrostatic force can be simplified with linear Poisson-Boltzmann approximation,

$$F_{EL} \approx 2\pi R \epsilon \epsilon_0 \psi_0^2 \kappa e^{-\kappa D} = 2\pi R q_f^2 e^{-\kappa D} / \kappa \epsilon \epsilon_0 \tag{2.8}$$

where the surface potential ψ_0 and surface charge density q_f are related by $q_f = \kappa \epsilon \epsilon_0 \psi_0$. For more general case, the surface charge density is calculated by using the Grahame equation [5]

$$q_f = \sqrt{8\epsilon_0 \epsilon k_B T I_{tot} N_A} \sinh\left(\frac{e_0 \psi_0}{2k_B T}\right) \tag{2.9}$$

where I_{tot} is the bulk total ionic strength, $I_{tot} = \frac{1}{2N_A} \sum_{i=1}^{n} \rho_{\infty i} z_i^2$.

It should be noted that the above equations are accurate for surface separation beyond one Debye length. At small separation one has to use numerical solutions of the Poisson-Boltzmann equation to obtain the exact interaction potential, for which there are no simple and accurate expressions. The charge regulation due to the counterion binding needs to be taken into account, therefore the strength of the double layer interaction is always less than that obtained at constant surface charge condition and higher than that at constant surface potential.

Combining the van der Waals force and the electrostatic double layer force, the DLVO force between two particles or two surfaces in a liquid can be expressed as:

$$F(D) = F_{vdW}(D) + F_{EL}(D) \tag{2.10}$$

In contrast to the double layer force, the van der Waals force is mostly insensitive to electrolyte strength and pH. Additionally, the van der Waals force is greater than the double layer force at small separation since it is a power law interaction, whereas the double layer force remains finite or increases much more slowly within the same separation range. The interplay between these two forces has many important consequences, thus understanding the individual forces and their contributions is a good way to control the stability of the colloidal suspensions.

2.1.3 Depletion Force

The depletion force exists in the systems containing particles with different length scales or particles and non-adsorbing polymer coils or micelles. In this dissertation, the depletion force arises between a micrometer-sized silica particle and a flat silica plate immersed in a dispersion of silica nanoparticles or surfactant micelles. The nanoparticles/micelles can be called depletion agents. Depletion agents are excluded from a shell of thickness of their radius around the larger silica particle (or silica plate), called the depletion zone. When two larger particles or surfaces are brought together and the distance h between two larger particle surfaces is less than diameter of depletion agents, $h < 2R$, their depletion zones will overlap and the depletion agents are expelled from the gap between the larger particles. The absence of depletion agents in the gap leads to a density gradient and an osmotic pressure causing the attractive depletion force between the larger particle surfaces. The range of the attraction is directly related to the radius of depletion agent, whereas the strength is proportional to the concentration of the depletion agent. Asakura and Oosawa [9, 10] first calculated the force per unit area between two parallel plates as being equal to the osmotic pressure of the surrounding depletion agent with simplest hard sphere approach:

$$\frac{F_{dep}}{A} = -\rho k_B T \Theta (2R - h), \quad h < 2R \tag{2.11}$$

where Θ is the Heaviside function. The depletion force depends on the particle number density ρ and absolute temperature T of the surrounding depletion agent. And $\frac{F_{dep}}{A} = 0$, for $h \geq 2R$ (Fig. 2.1).

A detailed insight into depletion forces is important for studying the stability and phase behavior of colloidal suspensions and for the understanding of properties of polymer-colloid mixtures and other self-assembling phenomena in liquid dispersions.

Fig. 2.1 When the distance between two surfaces is larger than the diameter of the depletion agents, $h \geq 2R$, the depletion agents can move into the gap and there is no depletion force acting on them. When the distance is small enough, $h < 2R$, the depletion agents are expelled from the gap and the net force acting on surfaces is equal to the pressure of the surrounding depletion agents

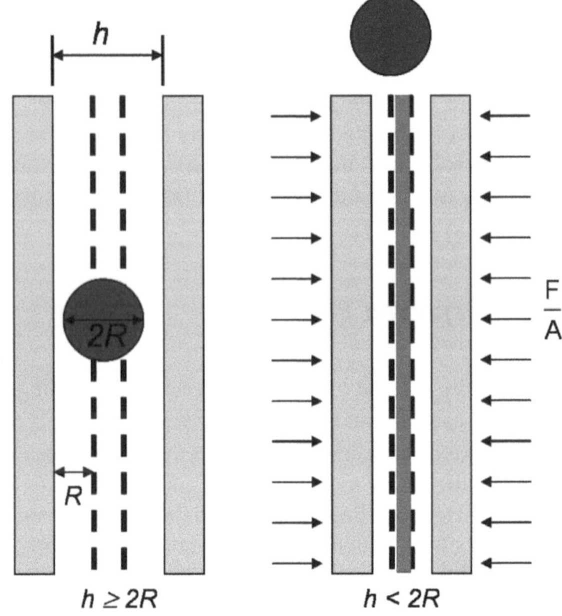

2.1.4 Structural force

Besides the depletion force, there is another non-DLVO force, called structural force or oscillatory force. It arises when the macroscopic surfaces are immersed in a relatively large concentration of the small colloidal particles. The structural force was first found by Israelachvili in the system of confined water molecules. The structural force is a generic feature also for colloidal particles, liquid crystals, and polyelectrolytes. These complex fluids can be considered as depletion agents for larger particles (surfaces). At large separation, the density of depletion agents in any highly restricted space is the same as that in bulk.

The density profile of depletion agents normal to a solid surface oscillates around the bulk density with a periodicity close to the distance of the depletion agent, and this oscillation extends several effective diameters of the depletion agent into the liquid. Within this range, the depletion agents are induced to order into quasi-discrete layers. When a neighboring surface approaches, the density distributions normal to both surfaces overlap and the depletion agents are squeezed out of the restricted space layer by layer so as to be accommodated between two surfaces. The variation of overlapping density profile with separation leads to an oscillating osmotic pressure.

The osmotic pressure as a function of separation is

$$P(h) = k_B T [\rho_s(h) - \rho_s(\infty)] \qquad (2.12)$$

where $\rho_s(h)$ is the density of depletion agent at two surfaces separated by a distance of h and $\rho_s(\infty)$ is the corresponding density at isolated surface. Thus an osmotic pressure arises once there is a change in the depletion agent's density at the surfaces as the surfaces approach each other. $\rho_s(h)$ is higher than $\rho_s(\infty)$ only when surface separations are multiples of the distance of the depletion agents and lower when at intermediate separations. At large separations, $\rho_s(h)$ approaches the value of $\rho_s(\infty)$, the osmotic pressure is zero.

As the last layer of depletion agent is squeezed out, $\rho_s(h \to 0) \to 0$, the osmotic pressure approaches a finite value given by

$$P(h \to 0) = -k_B T \rho_s(\infty) \tag{2.13}$$

which means the force at contact is negative. Equation 2.13 has the same form as the depletion force. Therefore, the depletion force is considered as a special case of the oscillatory force in the limit of very small separations.

The attractive surface-liquid interaction and geometric constraining both have influence on the variation of the layering of depletion agents with separation, the latter being essential because layering is still observed even in the absence of the attractive surface-liquid interaction.

For simple spherical depletion agents between two smooth surfaces the structural force is usually a decaying oscillatory function of distance (Fig. 2.2). For depletion agents with asymmetric shapes, the resulting structural force may also have a monotonically repulsive or attractive component. Structural forces depend not only on the properties of the depletion agent but also on the chemical and physical properties of the confining surfaces, such as the hydrophobicity, the morphology and the deformability of the surfaces.

In general, the oscillatory structural force consists of a harmonic oscillation coupled with an exponential decay function, thus it can be written as

$$\frac{F_{osc}(h)}{R} = A \exp\left(-\frac{h}{\xi_f}\right) \cos\left(2\pi \frac{h}{\lambda_f} + \theta_f\right) + \text{offset} \tag{2.14}$$

where R is the radius of the colloidal probe, and h is the separation between two confining surfaces. The three important parameters characterizing the oscillations are the amplitude A, the wavelength λ_f, and the decay length ξ_f. The amplitude describes the interaction strength of the particles, wavelength indicates the interparticle distance of the layered structuring, and decay length tells the degree of the ordering. This function is similar as the bulk pair correlation function which is valid at large interaction distance. Strictly speaking, this equation should apply at relatively large separation because the additional contribution of nonstructural forces exists at small distance.

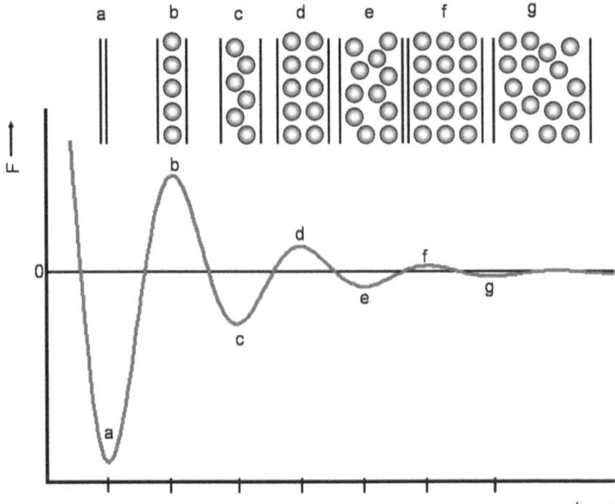

Fig. 2.2 Schematic illustration of the layering of spherical particles during the approach of two smooth surfaces and the corresponding measurable structural force. The density profile of particles changes with separation of two surfaces, resulting in different structuring of particles. At certain separation, particles are squeezed out of the gap to release the high inner pressure/force. The distance between two adjacent layers of particles relates to the inter-particle distance (illustration adopted from Israelachvili's book [5])

2.1.5 Hydrophobic Force

When using a hydrophobic surface, for example, replacing the solid substrate with a gas bubble or an other surface composed of hydrocarbons, the hydrophobic force plays a role in the system. The hydrophobic force describes the apparent repulsion between water and hydrophobic substances. Comparing to bond with hydrophilic molecules which have polar or ionic groups and hydrogen-bonding sites, water molecules have much less affinity to bond with the hydrophobic surface. The orientation of water molecules in contact with hydrophobic molecule is entropically unfavorable, therefore two hydrophobic molecules tend to come together and expel the water molecules into the bulk. This simple attractive force between hydrophobic molecules is favored because of the reduced total free energy of the system. Similarly, when water molecules are confined between hydrophobic surfaces, a net attractive force arises between the confining surfaces to expel the water molecules, which increases the translational and rotational entropy of water molecules and decreases the total free energy. Therefore, the attractive force, or hydrophobic force between hydrophobic surfaces is considered to be the consequence of water molecules migrating from a restricted space to the bulk water where there are unrestricted hydrogen-bonding opportunities and a lower free energy.

The attractive hydrophobic force is a long range force and much stronger than the van der Waals force [11–14]. It has a significant influence on the stability of colloids. Thus the force is needed to be taken into account when the system involves hydrophobic surfaces.

2.2 Theoretical Modelings

The developed theories are based on modeling by means of the integral equations of statistical mechanics [15], numerical simulations [16–18], and density-functional modeling [19–21]. As a rule, these approaches are related to complicated theoretical expressions or numerical procedures. The studies have shown that theoretical tools can describe oscillatory forces in a variety of model systems, such as hard spheres [16, 21, 22], polar fluids [23], liquid crystals [24], polyelectrolytes [25], and colloidal particles [17, 20]. In the case of nonionic micelles that can be modeled as hard spheres, Trokhymchuk et al. [22] proposed a quantitative analytical expression for the oscillatory force, which has been tested against both Monte-Carlo simulation data [22, 26] and data for stratifying free foam films [27].

2.2.1 Charged Particles

Based on Derjaguin-Landau-Verwey-Overbeck (DLVO) theory, a suspension of charged nanoparticles is modeled on an effective level which only includes the negatively charged silica nanoparticles explicitly [28–31]. Here, the electrostatic part of the DLVO potential is used [1]

$$u(r) = \tilde{Z}^2 e_0^2 \frac{\exp(-\kappa r)}{4\pi\epsilon_0\epsilon r} \tag{2.15}$$

where $Z = 4\pi(\sigma/2)^2 q_f$ is the total charge of a particle with diameter σ and surface charge density, q_f, which calculated from the Grahame Equation 2.9 with assuming the measured zeta potential ζ close to ψ_0.

In addition, $\tilde{Z}$ is the effective valency which is given by

$$\tilde{Z} = Z \exp(\kappa\sigma/2)/(1 + \kappa\sigma/2) \tag{2.16}$$

κ is the inverse Debye screening length, defined in Eq. 2.5 and can be also written as

$$\kappa = \sqrt{\frac{1}{\epsilon_0\epsilon k_B T}\left(\rho_c(z_c e_0)^2 + \sum_{k=1}^{K}\rho_k(z_k e_0)^2\right)} \tag{2.17}$$

where T is the temperature (set to 300 K), and ρ_c and z_c are the number density and valency of the counterions, respectively. The remaining sum refers to the additional salt ions. Assuming univalent counterions $|z_c| = 1$, the condition of charge neutrality between counterions and charged particles requires $\rho_c = |Z|\rho$. Equation 2.17 can then be rewritten as

$$\kappa = \sqrt{\frac{e_0^2}{\epsilon_0 \epsilon k_B T}(Z\rho + 2I_{salt}N_A)} \tag{2.18}$$

where $I_{salt} = \frac{1}{2N_A}\sum_{k=1}^{K} \rho_k(z_k)^2$ is the ionic strength of the additional salt, and N_A is Avogadro's constant.

For numerical reasons the soft sphere interaction $u_{SS}(r) = 4\epsilon_{SS}(\sigma/r)^{12}$ is added to the nanoparticles interaction which is, however, essentially negligible compared to the DLVO interaction energies at typical mean particle distances in this study.

Characteristic lengths as wavelength and decay length of bulk systems are extracted from bulk pair correlation functions $g_b(r)$. Within Monte Carlo (MC) simulations, $g_b(r)$ is determined using between 500 and 2000 particles depending on the volume fraction $\phi = (\pi/6)\rho\sigma^3$. Additionally, the integral equation for $g_b(r) \equiv h_b(r) + 1$ (with $h_b(r)$ being the total correlation function) consisting of the exact Ornstein-Zernike equation, $h(r_{12}) = c(r_{12}) + \rho \int dr^3 h(r_{13})c(r_{32})$ [32], combined with the approximate hypernetted chain (HNC) closure, $g(r) = \exp[-\beta u(r) + h(r) - c(r)]$ [33], is solved.

A convenient feature of using integral equations is that the asymptotic structure, that is, the dominant wavelength and correlation length of the function $h_b(r) = g_b(r) - 1$ in the limit $r \to \infty$ can be determined directly. This is done by analyzing the complex poles $q = \pm q_1 + iq_0$ of the structure factor $S_b(q) = 1 + \rho\tilde{h}_b(q)$ [19]. The pole with the smallest imaginary part determines the slowest exponential decay and thus yields an analytical description of the asymptotic behavior of $h_b(r)$ which reads

$$rh_b(r) = A_b exp(-q_0 r)cos(q_1 r - \theta_b), \quad r \to \infty \tag{2.19}$$

with $q_0 = \xi_b^{-1}$ playing the role of an inverse decay length (correlation length) and $q_1 = 2\pi/\lambda_b$ determining the wavelength of the oscillation. q_0 and q_1 can be also determined from the MC data by plotting the function $\ln(r\,|h_b(r)|)$. Wavelength and correlation length then follow from the oscillations and the slope of the straight line connecting the maxima at large r.

The HNC and MC results for wavelength in bulk λ_b as a function of the particle volume fraction ϕ are given in the main part of Fig. 2.3, showing that the two approaches are in good agreement. This is consistent with the observations previously reported [34] and justifies the use of HNC in the bulk system.

DFT [19, 20] predicts that, for sufficiently large h allowing a bulk-like region in the middle of the pore, the microscopic density profile should decay as $\rho(z) - \rho_b \to A_\rho exp(-q_0 z)cos(q_1 z - \theta_\rho)$, where q_0 and q_1 are exactly the same as in the bulk system at equal chemical potential μ (with bulk particle number density ρ_b), whereas the amplitude A_ρ and θ_ρ depend on the nature of fluid-wall (particle-confining surface) interactions. The same asymptotic behavior is expected for the so-called normal solvation pressure, $f(h) = P_{zz}(h) - P_{bulk}$.

The confined system modeled first by two infinite plane parallel, smooth, uncharged surfaces separated by a distance h along the z-direction [28, 30], fluid-wall (particle-confining surface) interaction is chosen to be purely repulsive and read as

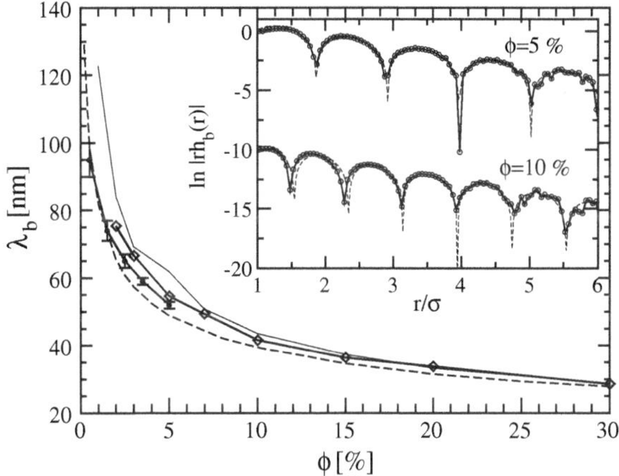

Fig. 2.3 Dominant wavelength λ_b characterizing $h_b(r) = g_b(r) - 1$ as a function of the volume fraction according to HNC (*solid line*) and MC (*diamonds*). Also shown are the HNC data for λ_s [28] (*dashed line*) and the corresponding SANS data for λ_s [30] (*filled circles*, with error bars). The inset shows two MC results for the function $\ln(r\,|h_b(r)|)$ (*circles*). The asymptotic fit functions are plotted as *dashed lines*

$$u_{\mathrm{FS}}^{\mathrm{SW}}(z) = \frac{4}{5}\pi\epsilon_{\mathrm{w}}\left(\frac{\sigma}{z}\right)^{9} \tag{2.20}$$

The MC simulation in the grand canonical (GC) ensemble can be employed to investigate the confined systems, that is, at constant temperature, wall separation, box area parallel to the walls, and constant chemical potential μ [28, 35]. Furthermore, the inverse Debye length should be fixed at the value corresponding to the bulk volume fraction. The solvation pressure $f(h) = P_{zz}(h) - P_b$ with P_b being the bulk pressure and $P_{zz}(h)$ the normal component of the pressure tensor [35] exhibits oscillations with varying surface separation. This quantity is related to the oscillatory forces of the AFM experiment via Derjaguin's approximation [21]. The functions $f(h)$ are fitted according to the expression

$$\tilde{f}(h) = A_{\mathrm{f}}\exp\left(-\frac{h}{\xi_{\mathrm{f}}}\right)\cos\left(\frac{2\pi h}{\lambda_{\mathrm{f}}} - \theta_{\mathrm{f}}\right) \tag{2.21}$$

with λ_f and ξ_f being the wavelength and decay length, respectively.

In the whole particle concentration range considered, the oscillatory asymptotic decay of $f(h)$ (determined by a wavelength λ_f) is indeed well described by the leading bulk wavelength (and correlation length), implying $\lambda_f = \lambda_b$. This is particularly well-demonstrated by the logarithmic representation in the inset of Fig. 2.4. Thus, the GCMC simulation results for the charged silica particles confirm the DFT predictions. As Eq. 2.19 derived from Ornstein-Zernike theory Eq. 2.21 is only valid for $h \to \infty$. However the asymptotic expression is found to provide a good approximation of

Fig. 2.4 Two examples of
the solvation pressure *f(h)* as
obtained by GCMC (*filled
circles*) together with the
asymptotic fits (*solid line*)
obtained with the bulk values
of $q_1 = 2\pi/\lambda_b$ and q_0.
Included are the resulting
structural forces $F(h)/2\pi R$
(*dashed line*). The inset shows
a logarithmic plot of *f(h)*

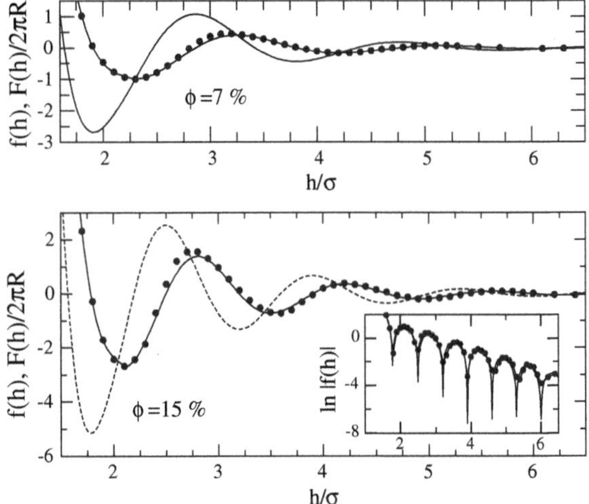

the oscillations already at remarkably small wall separations. The full curve is well
described by the asymptotic formula already after the first minimum at $h = h_{min}$.
For smaller separations $h \leq h_{min}$, a cubic polynomial fit is used:

$$\tilde{f}(h) = a_0 + a_1 h + a_2 h^2 + a_3 h^3 \tag{2.22}$$

The coefficients are adjusted such that both the pressure and its derivative are
continuous at h_{min}. Then, an accurate fit formula for *f(h)* can be found by immediately
integrating *f(h)* [21] to obtain the solvation force $F(h)/2\pi R$, results for which are
included in Fig. 2.4.

In order to obtain the impact of surface potential on the other quantities of the
solvation forces, i.e. amplitude and phase shift, which is predicted by the DFT, the
confined system is modeled by two infinite plane parallel, smooth, charged surfaces
separated by a distance h along the z-direction [31]. Their surface potential mimics
the experimental conditions. A simplest version of linearized Poisson-Boltzmann
(PB) theory is first employed

$$u_I(z) = W_S \exp\left(-\kappa\left(\frac{h}{2} - z\right)\right) + W_S \exp\left(-\kappa\left(\frac{h}{2} + z\right)\right) \tag{2.23}$$

The decay of potential is determined by the bulk Debye screening length κ and the
wall charge only come into play through a prefactor $W_S = 64\pi\epsilon_0\epsilon\gamma_f\gamma_s R \left(k_B T/e_0\right)^2$,
where $\gamma_{F/S} = \tanh\left(e_0\psi_{F/S}/4k_B T\right)$, $\psi_{F/S}$ is the surface potential of the fluid particles
(F) and the solid walls *(S)*, and R is the radius of the fluid particle, $R = \sigma/2$.
This model has the consequence that the oscillations of the normal pressure become
weakened with increasing $|\psi_S|$, which is in contradiction to the experiment.

Thus a modified fluid-wall potential is developed. It takes additional wall counterions into account which change the screening between charged walls and colloidal particles. The expression of a silica ion and one of the charged walls is read as

$$u_{\text{FS}}^{\text{LSA}}(z) = 64\pi\epsilon_0\epsilon\gamma_{\text{F}}\gamma_{\text{S}}R\left(\frac{k_BT}{e_0}\right)^2 \exp\left[-\kappa_{\text{W}}(z-R)\right] \tag{2.24}$$

where the screening parameter depends on ψ_S and is space-dependent,

$$\kappa_{\text{W}}(z) = \sqrt{\frac{e_0^2}{\epsilon_0\epsilon k_BT}\left(Z\rho + 2I_{salt}N_{\text{A}} + \frac{|q_{\text{S}}|}{e_0z}\right)} \tag{2.25}$$

and the confining surface charge density q_{S} is related to the surface potential via the Grahame equation.

The total fluid-wall interaction is therefore given by

$$u_{\text{FS}}(z) = u_{\text{FS}}^{\text{LSA}}(h/2-z) + u_{\text{FS}}^{\text{LSA}}(h/2+z) + u_{\text{FS}}^{\text{SW}}(h/2-z) + u_{\text{FS}}^{\text{SW}}(h/2+z) \tag{2.26}$$

where $u_{\text{FS}}^{\text{LSA}}$ and $u_{\text{FS}}^{\text{SW}}$ are given by Eqs. 2.24 and 2.20, respectively. This induces a non-monotonic behavior of the fluid-wall interaction potential as a function of the wall charge.

2.2.2 Nonionic Surfactant Micelles

Using the Derjaguin's approximation, one can express the surface force, F, between a spherical particle and a planar plate in the form:

$$F(h) = 2\pi RW(h) \tag{2.27}$$

where R is the particle radius; h is the surface-to-surface distance between the particle and the plate; $W(h)$ is the interaction energy per unit area of a plane-parallel liquid film of thickness h. In the considered case of nonionic surfactant micelles, W can be expressed as a sum of contributions from the van der Waals forces, W_{vdW}, and oscillatory structural forces due to the surfactant micelles, W_{osc} [5, 27]

$$W(h) = W_{osc} + W_{vdW} = W_{osc} - \frac{A_H}{12\pi h^2} \tag{2.28}$$

The surface charge of the confining surfaces can be neglected, since confining surface charge would not change the oscillation wavelength and decay length, but increase the amplitude w_0 [20].

The combination of Eqs. 2.27 and 2.28 yields

$$F = 2\rho R \frac{k_B T}{d^2} \left(\frac{W_{osc} d^2}{k_B T} - \frac{A_H}{12\pi (h/d)^2 k_B T} \right)$$
(2.29)

where d is the micelle diameter; h/d is the dimensionless surface-to-surface distance. Furthermore, the expression for W_{osc} due to Trokhymchuk et al. [22] is used:

$$\frac{W_{osc} d^2}{k_B T} = -\frac{p_{hs} d^3}{k_B T}(1 - h/d) - \frac{2\sigma_{hs} d^2}{k_B T}, \quad 0 \le h/d < 1$$
(2.30)

$$\frac{W_{osc} d^2}{k_B T} = w_0 \cos(\omega h/d + \varphi_1) \exp^{-qh/d} + w_1 \exp^{\delta(1-h/d)}, \quad h/d \ge 1$$
(2.31)

where p_{hs} is the pressure of a hard-sphere fluid expressed through the Carnahan-Starling formula [36], and σ_{hs} is the scaled-particle-theory [37] expression for the excess surface free energy of a hard-sphere fluid:

$$\frac{p_{hs} d^3}{k_B T} = \frac{6}{\pi} \phi \frac{1 + \phi + \phi^2 - \phi^3}{(1 - \phi)^3}$$
(2.32)

$$\frac{\sigma_{hs} d^2}{k_B T} = -\frac{9}{2\pi} \phi^2 \frac{1 + \phi}{(1 - \phi)^3}$$
(2.33)

The parameters w_0, ω, φ_1, q, w_1 and δ in Eq. 2.31 depend on the hard-sphere (micelle) volume fraction, ϕ, as follows [22]

$$w_0 = 0.57909 + 0.83439\phi + 8.65315\phi^2$$
(2.34)

$$\omega = 4.45160 + 7.10586\phi - 8.30671\phi^2 + 8.29751\phi^3$$
(2.35)

$$q = 4.78366 - 19.64378\phi + 37.37944\phi^2 - 30.59647\phi^3$$
(2.36)

$$w_1 = \frac{2\sigma_{hs} d^2}{k_B T} - w_0 \cos(\omega + \varphi_1) \exp(-q)$$
(2.37)

$$\varphi_1 = 0.40095 + 2.10336\phi, \quad \delta = \frac{\pi_1}{w_1}$$
(2.38)

where

$$\pi_1 = \frac{6}{\pi}\phi \exp\left(\frac{\Delta\mu_{hs}}{k_B T}\right) - \frac{p_{hs} d^3}{k_B T} - \pi_0 \cos(\omega + \varphi_2)\exp(-q) \tag{2.39}$$

$$\frac{\mu_{hs}}{k_B T} = \phi\frac{8 - 9\phi + 3\phi^2}{(1 - \phi)^3} \tag{2.40}$$

$$\pi_0 = 4.06281 - 3.10572\phi + 76.67381\phi^2 \tag{2.41}$$

$$\varphi_2 = -0.39687 - 0.3948\phi + 2.3027\phi^2 \tag{2.42}$$

The parameters w_0, ω, and q defined by Eqs. 2.34–2.36 characterize, respectively, the amplitude, wavelength and decay length of the oscillations (see Eq. 2.31). The last term in Eq. 2.31 ensures the correct height of the first (the highest) maximum [22]. Note that for a hard sphere fluid, the amplitude, wavelength and decay length of the oscillations depend on the particle volume fraction, ϕ, in accordance with Eqs. 2.34–2.36.

Equation 2.29, along with Eqs. 2.30–2.42, determines the theoretical dependence $F(h, \phi)$ at given colloidal probe radius, R, and micelle diameter, d. In particular, for a given micelle volume fraction, ϕ, p_{hs}, σ_{hs}, w_0, ω, q, μ_{hs}, π_0 can be first calculated, and φ_2 from Eqs. 2.32–2.36 and Eqs. 2.40–2.42; after that, w_1, φ_1, π_1 and δ can be calculated from Eqs. 2.37–2.39; next W_{osc} is computed from Eqs. 2.30–2.31, and finally F, from Eq. 2.29.

The fitting procedure is as follows. The experimental force, F_{exp} is given as a function of the experimental distance, $h_{exp} = h + \Delta h$, where h is the theoretical distance and Δh is the difference between the positions of the experimental and theoretical coordinate origins on the h-axis. The fitting by means of the least-squares method consists of numerical minimization of the following merit function:

$$\Phi(\Delta h, \phi) = \sum_i \left[F\left(h^i_{\exp} - \Delta h, \phi\right) - F^i_{\exp}\left(h^i_{\exp}\right)\right]^2 \tag{2.43}$$

where $F^i_{\exp}\left(h^i_{\exp}\right)$ is the set of experimental data numbered by the index i, and the summation is carried out over all experimental points. It is important to note that in the fitting procedure, the points from the non-equilibrium portions of the experimental curves have to be excluded, because the theoretical curve gives the equilibrium force-versus-distance dependence.

When ϕ is known, the variation of Δh is equivalent to a simple horizontal translation of the experimental curve with respect to the theoretical one, the latter being uniquely determined. The minimization of Φ with respect to Δh corresponds to the best coincidence of the two curves. When ϕ is not known, both Δh and ϕ should be varied to minimize numerically Φ in Eq. 2.43, and to find the best fit. When calculating the theoretical curves, in Eq. 2.29 the value $A_H = 7 \times 10^{-21}$ J of the Hamaker constant for silica/water/silica films is used [5]. The effect of van der Waals forces is essential only at the lowest investigated micellar concentrations, where the oscillatory amplitude w_0 is relatively small.

References

1. Verwey, E. J. W., & Overbeek, J. T. G. (1948). *Theory of stability of lyophobic colloids.* Amsterdam: ELSEVIER.
2. Derjaguin, B. V., & Landau, L. (1941). *Acta Physicochim URSS, 14,* 633.
3. Hamaker, H. C. (1937). *Physica, 4,* 1058–72.
4. Lifshitz, E. M. (1956). *Soviet Physics JETP USSR, 2,* 73–83.
5. Israelachvili, J. N. (1992). *Intermolecular and surface forces.* London: Academic Press.
6. Russel, W., Saville, D., & Schowalter, W. (1989). *Colloidal dispersions.* Cambridge, UK: Cambridge University Press.
7. Hutter, R. J. (2002). *Foundations of colloid science.* Oxford: Oxford University Press.
8. Derjaguin, B. V. (1934). *Kolloid Z, 69,* 155–64.
9. Asakura, S., & Oosawa, F. (1954). *Journal of Chemical Physics, 22,* 1255–1256.
10. Asakura, S., & Oosawa, F. (1958). *Journal of Polymer Science, 33,* 183–192.
11. Israelachvili, J., & Pashley, R. (1982). *Nature, 300,* 341–342.
12. Israelachvili, J., & Pashley, R. (1984). *Journal of Colloid and Interface Science, 98,* 500–514.
13. Rabinovich, Y., & Derjaguin, B. (1988). *Colloids and Surfaces, 30,* 243–251.
14. Claesson, P., & Christenson, H. (1988). *Journal of Physics and Chemistry, 92,* 1650–1655.
15. Attard, P., & Parker, J. (1992). *Journal of Physics and Chemistry, 96,* 5086–5093.
16. Blawzdziewicz, J., & Wajnryb, E. (2005). *Europhysics Letters, 71,* 269–275.
17. Trokhynichuk, A., Henderson, D., Nikolov, A., & Wasan, D. (2005). *Langmuir, 21,* 10240–10250.
18. Schoen, M., & Klapp, S. H. L. (2007). *Nanoconfined fluids. Soft matter between two and three dimensions.* New York: Wiley.
19. Evans, R., Henderson, J., Hoyle, D., Parry, A., & Sabeur, Z. (1993). *Molecular Physics, 80,* 755–775.
20. Grodon, C., Dijkstra, M., Evans, R., & Roth, R. (2005). *Molecular Physics, 103,* 3009–3023.
21. Gotzelmann, B., Evans, R., & Dietrich, S. (1998). *Physical Review E, 57,* 6785–6800.
22. Trokhymchuk, A., Henderson, D., Nikolov, A., & Wasan, D. (2001). *Langmuir, 17,* 4940–4947.
23. Klapp, S., & Schoen, M. (2002). *Journal of Chemical Physics, 117,* 8050–8062.
24. Schoen, M., Gruhn, T., & Diestler, D. (1998). *Journal of Chemical Physics, 109,* 301–311.
25. Jonsson, B., Broukhno, A., Forsman, J., & Akesson, T. (2003). *Langmuir, 19,* 9914–9922.
26. Kralchevsky, P., & Denkov, N. (1995). *Chemical Physics Letters, 240,* 385–392.
27. Basheva, E. S., Kralchevsky, P. A., Danov, K. D., Ananthapadmanabhan, K. P., & Lips, A. (2007). *Physical Chemistry Chemical Physics, 9,* 5183–5198.
28. Klapp, S. H. L., Qu, D., & von Klitzing, R. J. (2007). *Journal of Physical Chemistry B, 111,* 1296–1303.
29. Klapp, S. H. L., Grandner, S., Zeng, Y., & von Klitzing, R. (2008). *Journal of Physics: Condensed Matter, 20,* 494232.
30. Klapp, S. H. L., Zeng, Y., Qu, D., & von Klitzing, R. (2008). *Physical Review Letters, 100,* 118303.
31. Grandner, S., Zeng, Y., von Klitzing, R., & Klapp, S. H. L. (2009). *Journal of Chemical Physics, 131,* 154702.

32. Ornstein, L., & Zernike, F. (1914). *Proceedings of the Academy of Sciences Amsterdam, 17*, 793.
33. Hansen, I. R., & McDonald, J. P. (2006). *Theory of simple liquids* (3rd ed.). Amsterdam: Academic Press.
34. Hopkins, P., Archer, A., & Evans, R. (2005). *Physical Review E, 71*, 027401.
35. Schoen, M., Klapp, S. H. L (2007). *Reviews in computational chemsitry* (Vol. 24). New Jersey: WILEY-VCH.
36. Carnahan, N., & Starling, K. (1969). *Journal of Chemical Physics, 51*, 635.
37. Reiss, H., Frisch, H., Helfand, E., & Lebowitz, J. (1960). *Journal of Chemical Physics, 32*, 119–124.

Chapter 3
Experimental Section

3.1 Preparation of Materials

3.1.1 Colloidal Nanoparticle Suspensions

Ludox grade colloidal silica nanoparticle suspensions, named TMA 34 (deionized), HS 40 (stabilized with Na^+), and SM 30 (stabilized with Na^+) were purchased from Aldrich (Taufkirchen, Germany). The original stock of colloidal suspensions was dialyzed with Milli-Q water (Millipore, Billerica, MA, USA) for two weeks. The dialysis tubes (Aldrich, Germany) with pore size of 1000 MWCO were used to remove any remaining ions and ionic contaminants. After dialysis, particle suspensions of varying concentrations were prepared with Milli-Q water as solvent. The weight percentage, c, and density of the solutions, ρ_s, were determined by weighing a known volume of the sample before and after drying (24 h at 400 °C). The volume fraction was converted from the weight percentage as $\phi = \frac{c \times \rho_s}{\rho_p}$. The mass density of silica particle ρ_p was set as manufactural value of $2.2\,\mathrm{g\,mL^{-1}}$. The mass density of suspension had a dependency on the particle concentration as in Fig. 3.1. The silica nanoparticle suspensions have a pH of about 6.5. Whenever needed, sodium chloride NaCl (suprapur 99,99 %, Merck) was used to tune the ionic strength of the suspensions. Ludox silica suspensions have been stored in plastic tubes.

3.1.2 Surfactant Solutions

In terms of the measurements on deformable bubble surfaces, surfactants were used to tune the surface tension. β-dodecylmaltoside (β-$C_{12}G_2$) (Glycon Biochemicals, Luckenwalde), sodium dodecyl sulfate (SDS) (Sigma-Aldrich, purity $>99.9\,\%$) and hexadecyltrimethylammonium bromide ($C_{16}TAB$) (Sigma-Aldrich, purity $>99\,\%$) were used as received. The critical micelle concentrations (CMC) are 0.17 mM,

Y. Zeng, *Colloidal Dispersions Under Slit-Pore Confinement*, Springer Theses,   23
DOI: 10.1007/978-3-642-34991-1_3, © Springer-Verlag Berlin Heidelberg 2012

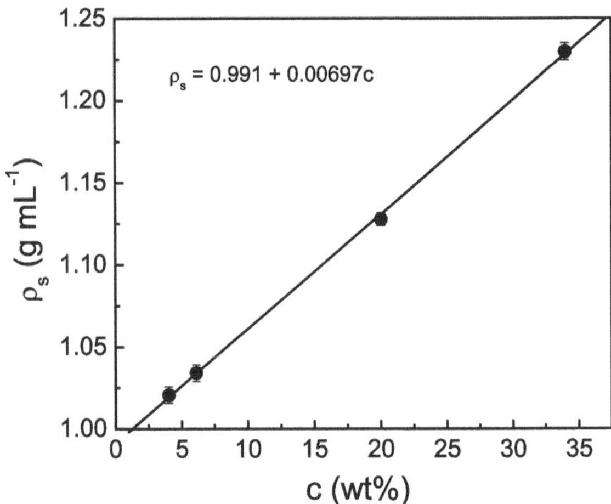

Fig. 3.1 The dependency of mass density of the particle suspensions on the particle concentration, which is needed to convert weight percentage to volume fraction

8 mM, and 1 mM for β-$C_{12}G_2$, SDS, and C_{16}TAB, respectively. All surfactant solutions were prepared with Milli-Q water. The surfactant concentrations were always well below the CMC.

3.1.3 Micelle Solutions

The nonionic surfactants, polyoxyethylene lauryl ether (Brij 35), and polyoxyethylene sorbitan monolaurate (Tween 20), products of Sigma, were used without further purification. The molecular weight of Brij 35 is $1198\,\mathrm{g\,mol}^{-1}$; its critical micellization concentration (CMC) is $9 \times 10^{-5}\,\mathrm{M}$, and the micelle diameter is $d = 8.8\,\mathrm{nm}$. The micelles are spherical up to 150 mM Brij 35 concentration, but they undergo a transition to elongated micelles at higher concentrations. The molecular mass of Tween 20 is $1225\,\mathrm{g\,mol}^{-1}$; its CMC is $4.9 \times 10^{-5}\,\mathrm{M}$, and the micelle diameter is $d = 7.2\,\mathrm{nm}$.

3.1.4 Polyelectrolytes Solutions

Polyethylenimin (PEI), 750 kDa, 50 wt% solution in water, poly(allylamine hydrochloride) (PAH), 65 kDa, poly(sodium 4-styrenesulfonate) (PSS), 70 kDa, and poly(acrylic acid) (PAA), 450 kDa were purchased from Sigma Aldrich (Germany). Hyaluronic acid (HA) in the sodium hyaluronate form, 150 kDa, was obtained from Amersham Bioscience (Munich, Germany). The polyelectrolytes solutions were prepared by dissolving the amount of polymer corresponding to a concentration of 10^{-2} monoM in Milli-Q water. Certain amount of sodium chloride (suprapur

99,99 %, Merck) was added afterwards when necessary. In order to overcome solubility problems, a solution of $1\,mg\,mL^{-1}$ of HA was used.

3.2 Preparation of Different Surfaces

3.2.1 Silica Surface

Silica, as colloids or a plate, is an amorphous material which is often used as model system for studying the surface chemistry and interaction of many systems. The surface of silica is well known to be negatively charged due to the ionization of silanol groups in contact with water: $SiOH + H_2O \rightleftharpoons SiO^- + H_3O^+$. The surface charge of silica varies with pH, electrolyte concentration, and cleaning process. It has been proposed in the literature that a gel-like surface layer forms when silica is in contact with water. This gel-like layer is about 2–6 nm thick and composed of silanol and silicic acid groups. This hypothesis may explain the high surface charge and low potentials of the silica surface, and also the additional non-DLVO repulsion at small separation due to the steric repulsion between two overlapped layers.

Silicon wafers (Wacker Siltronic AG, Germany) were cleaned in a 1:1 mixture of piranha solution (H_2O_2/H_2SO_4 solution) for 30 min followed by extensive rinsing with Milli-Q water. Afterwards, the etched silicon wafers were stored under Milli-Q water in a glass container. The above procedure yielded a fully hydrophilic surface. Just before the experiment, the substrate was taken out of water and dried in a nitrogen flux.

3.2.2 Mica Surface

Mica is easily cleaved into atomically smooth layers and hence is widely used as an important substrate in many fundamental studies. The highly perfect cleavage is explained by the hexagonal sheet-like arrangement of its atoms. Mica is a layered dioctahedral aluminosilicate represented as $KAl_2(AlSi_3)O_{10}(OH)_2$. Each mica sheet consists two silicate layers joined together by aluminum atoms. The substitution of aluminum for silicon in the silicate layers results in a net negatively charged lattice, which is neutralized by potassium ions present between aluminosilicate sheets. Therefore mica can be cleaved along the plane of potassium ions. In air the mica surface is neutralized while in water it acquires a high negative surface charge by dissociation and ion exchange. The apparent surface potential of mica in pure water is found to be $-160\,mV$, which is decreased slightly as the ionic strength of the solution increases. The fresh mica surface was prepared by cleaving the mica sheets with tweezer, and then deposited on top of a silicon wafer.

3.2.3 Hydrophobic Substrate

In order to avoid the strong capillary force between the AFM silica probe and hydrophilic substrate in air during spring constant determination, a modified silicon wafer with contact angle >100 °C is needed. 50 µL of heptadecafluoro-1,1,2,2-tetrahydrodecyl simethylchlorosilane (ABCR, Karlsruhe) was pipette in a small glass vial, which could be closed very tightly. A clean silicon wafer (1–3 cm^2) was placed inside of the glass vial without that the wafer was in contact with the silane. After closing the vial, the silicon wafer was leaving inside for 24 h at room temperature. The modified silicon wafer was then taken out and heated in an oven to 70 °C for 10–20 min in order to remove non-bounded silane. The contact angle of the hydrophobic silicon wafer was measured >100 °C.

3.2.4 Polyelectrolyte Layer-by-Layer Adsorption on Surfaces

On silicon wafer: After cleaning in a 1:1 mixture of H_2O_2/H_2SO_4 solution for 30 min and then extensively rinsing in Milli-Q water, the silicon wafers were dipped into an aqueous solution of 10^{-2} monoM PEI for 30 min and then rinsed gently in Milli-Q water. The layer-by-layer self assembly of polyelectrolytes consists of sequential dipping of silica substrate into polyanion and polycation solutions [1]. The adsorption time for each layer was 20 min, after which the substrate was rinsed by dipping the wafers three times into fresh Milli-Q water for 1 min to remove any loosely bound polyelectrolytes. This process was repeated a defined n times to obtain a multilayer consisting of (polyanion/polycation)$_n$ layers. The multilayers were not dried between different deposition steps. After the last adsorption step, the samples were dried with nitrogen stream and stored in clean glass vessels.

On silica sphere: The polyelectrolyte multilayers were accomplished by adsorption from polyelectrolyte solutions according to the previous description [2]. 500 mL of a 10^{-2} monoM PEI solution was added to 2.5 mL of 6.7 µm sized silica suspension. The samples were sonicated and the adsorption solution was left to stand for a minimum time of 30 min. The solution was then centrifuged at 4300 rpm for 20 min and the supernatant was removed. 500 mL of water was added to the sample and the solution was sonicated and left to stand for 20 min. A total of three washing cycles were performed, after the adsorption of each polyelectrolyte layer. To the remaining colloidal solution was alternately added 500 mL of a 10^{-2} monoM PSS and PAH solution. Similar adsorption and washing steps were performed. Successful deposition was verified by ζ-potential measurements after each deposit step.

3.3 Methods

3.3.1 Atomic Force Microscopy

The atomic force microscopy (AFM) was invented by Binning et al. [3] and first used in imaging the topography and morphology of samples at different resolution. Later on it was modified by Ducker et al. [4–6] for use in measuring interactions between different objects. The main principle of AFM is to reflect a laser beam on the free end of a cantilever. The reflected laser beam is measured by a detector. Any positional changes of the cantilever like bending or twisting are recorded by the detector. A scanner which made of piezoelectric materials is used to move the cantilever with high resolution in all directions. The cantilever sensitivity can be varied by choosing different spring constants in the range of $0.002–400\,\mathrm{N\,m^{-1}}$. The probe is mounted to the free end of cantilever. Normally for imaging the probe is mounted as a sharp tip, while for interaction measurements it is a glued micrometer-sized silica sphere. The AFM setup is depicted in Fig. 3.2.

3.3.1.1 Imaging

Two major imaging modes are normally used in the experiments: contact mode and tapping mode. In contact mode the probe and the sample are in contact during

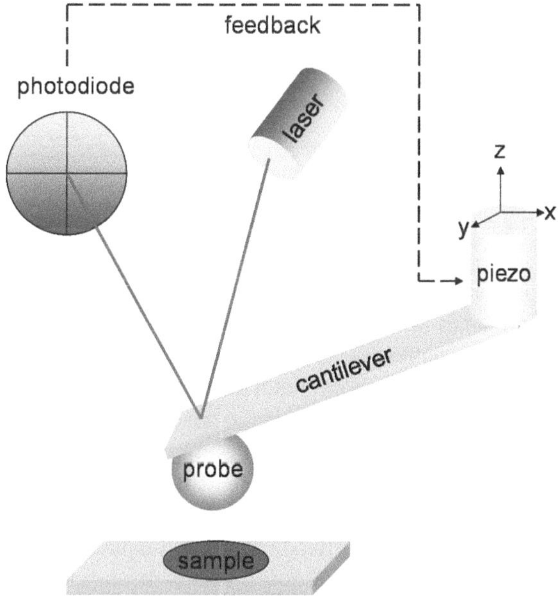

Fig. 3.2 Schematic AFM setup. Four main parts are included: laser beam, photodiode detector, cantilever with a probe and piezoelectric scanners to move the cantilever in three directions. For AFM imaging, a cantilever with sharp tip is used and feedback electronics are required to maintain either the force or amplitude during scanning

imaging. Optimally applied force between the probe and sample needs to be selected via a set point for the vertical detector response. During scanning the set point is kept constant. If a peak/valley is reached on the sample when the probe passes by, the cantilever will push upwards/downwards as a reaction but the feedback will raise/lower the cantilever to maintain the detector signal constant. By recording the cantilever height for each sample position, one obtains a three-dimensional image of the sample surface topography.

Tapping mode, or AC mode, is a common mode for imaging samples, especially in liquid. The cantilever oscillates close to its resonant frequency during scanning. The system attempts to keep the amplitude of the oscillation constant by using a feedback on the height to raise or lower the probe if there are any irregularities on the sample. Atomic force microscope MFP-3D provides a parallel imaging method via iDrive^{TM} (Asylum Research, CA, USA), which uses Lorentz force to magnetically actuate a cantilever with an oscillating current that flows through a v-shaped cantilever instead of acoustic piezo-driven placed close to the cantilever in the commercial AFM. IDrive can eliminate the multitude of resonance peaks due to the mechanical coupling of the piezo to the cantilever and liquid. A resonant peak can be easily defined with auto tune in aqueous media.

The morphology of polyelectrolytes-coated silicon wafer was performed in tapping mode with iDrive cantilever BL-TR400PB (Asylum Research, Santa Barbara, CA) in Milli-Q water via a MFP-3D setup produced by Asylum Research, Inc. and distributed by Atomic Force (Mannheim, Germany). The root mean square (RMS) roughness of polyelectrolyte layers was calculated from height images with bound software in each $1 \times 1 \mu m^2$ box of the image so as to be comparable to standard neutron reflectivity measurements with $1\,\mu m$ beam correlation length. The final value is an average of those calculated at different positions on each image.

$$R_{RMS} = \sqrt{\frac{1}{n}\sum_{i}^{n} y_i^2} \tag{3.1}$$

3.3.1.2 Force Measurement

The force between the probe and sample can be recorded as the AFM probe approaches and retracts from the sample surfaces. What is directly measured is a profile of deflection (volts) versus ZSensor displacement (μm). During approach the probe will at some point be in contact with the sample. If further approach is attempted, the deflection of the cantilever linearly increases with the ZSensor displacement. This linearly increasing part is called the constant compliance region, and the slope of deflection versus displacement in this region is referred to as the deflection sensitivity. Note the deflection sensitivity must be determined on a rigid surface, where the drive displacement equal to the bending amount. The algorithm used to convert the deflection versus displacement data into force versus apparent separation is illustrated in Fig. 3.3 as in the protocol of Ducker et al. [6].

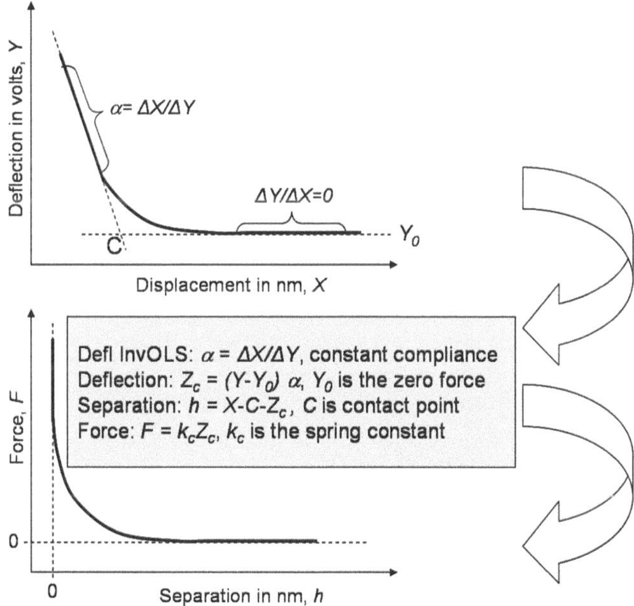

Fig. 3.3 The algorithm to convert primary data into force versus separation. For solid surfaces the slope of the force curve become infinity at zero separation (i.e. in contact)

The deflection inverse optical lever sensitivity (InvOLS) α in the constant compliance region is defined as

$$\alpha = \frac{\Delta X}{\Delta Y} \tag{3.2}$$

where X and Y are the piezo displacement in meters and deflection signal in volts, respectively. The voltage signal of the deflection thus can be related to the bending in meters, Z_c, by

$$Z_c = (Y - Y_0)\alpha \tag{3.3}$$

where Y_0 is the deflection at infinite displacement. This subtraction is needed based on the assumption that there is no interaction between AFM probe and surface at large separation. The separation between the probe and surfaces is calculated as

$$h = (X - C) - Z_c \tag{3.4}$$

where constant C is the displacement value at "contact point". The "contact point" is taken to be the point at which the linear compliance line reaches zero force. The force F then can be calculated through the Hooke's law

$$F = k_c Z_c \tag{3.5}$$

where k_c is the spring constant of the cantilever, which was determined with a thermal noise power spectra before or after the experiment [7] with a hydrophobic substrate and yielded values in the range 0.01–$0.08\,\mathrm{N\,m^{-1}}$ (see Sect. 3.2.3).

A silica sphere (Bangslabs, USA) with radius of $R = 3.35\,\mu\mathrm{m}$ was glued with epoxy glue (UHU Endfest Plus 300) at the end of a tipless rectangular cantilever (CSC12, MikroMasch, Estonia) using a three-dimensional microtranslation stage according to the previous procedure [4]. Immediately before each experiment the silica sphere with cantilever was cleaned by exposure to a plasma cleaner for 20 min to remove all the organic contaminants and to create a high density of hydrophilic silanol groups (Si-OH) on the surface.

The cantilever was placed into a cantilever holder and the particle probe was positioned roughly a few μm above the substrate. Then few drops of the target sample solution was put onto the substrate, and the probing head was fully immersed in the solution. Force-separation curves were collected with a MFP-3D. No adsorption of silica nanoparticles on the AFM probe or substrates is expected because silica and mica surfaces are negatively charged in the experimental conditions (pH $\approx$ 6–7), while the non-ionic surfactant is partially adsorbed at the surface.

The optimal scanning velocity was in the range of 150–$400\,\mathrm{nm\,s^{-1}}$ for silica nanoparticle suspensions and 5–$100\,\mathrm{nm\,s^{-1}}$ for non-ionic surfactant solution, respectively, over a scan size of 300–$400\,\mathrm{nm}$. Chan and Engel showed that hydrodynamic drainage forces were negligible at these approach speeds [8, 9]. For each sample solution, altogether 30–40 force-distance curves were recorded at the same lateral position (usually at the centre). To quantitatively study the structuring of nanoparticles, the oscillatory forces are fitted with Eq. 2.14. As the silica microsphere is $6.7\,\mu\mathrm{m}$ in diameter, by Derjaguin approximation the silica probe surface can be considered as a flat surface because of the comparatively small force distance (<300 nm). Thus force per probe radius, $\frac{F(h)}{R}$, is the measure of interaction energy per area. All experimental force curves were fitted with Eq. 2.14. Beside the three mentioned parameters a phase shift (θ_f) and a force offset (offset) also had to be fitted.

3.3.1.3 Surface Elasticity Measurements

First, it is necessary to calibrate the deflection inverse optical lever sensitivity (InvOLS). The previous cases have involved solid substrates that are much stiffer than the cantilevers. The deflection InvOLS can thus be simply calibrated by finding the slope of deflection versus ZSensor once the surfaces are in contact. In the present case the cantilever and bubble can have similar stiffnesses so that the calibration must be done separately, before or after the force measurements, by pressing the particle against a rigid surface. The AFM photodiode voltage was converted to cantilever deflection using the detector sensitivity determined before the experiment and then converted into force via Eq. 3.5.

The conversion from ZSensor position to actual particle-bubble separation is more complicated. The nominal separation is defined as in Eq. 3.4. This definition does not consider deformation, so for rigid surfaces the nominal separation coincides with the actual separation. Then for deformable surfaces, the actual separation is the nominal separation minus the deformation

$$\Delta h = \Delta X - \Delta \delta \tag{3.6}$$

An attractive force between AFM probe and substrate causes an extension and a positive deformation while a repulsive force causes a negative deformation. During the measurements, an absolute measure of the shape change of the bubble surface is not known, only the changes in ΔX are measured. This problem cannot be resolved without measurement of actual particle-bubble separation using an independent method, e.g. interferometry. Thus it is difficult to plot F versus h. Instead F versus ΔX is presented in this thesis.

$$\Delta X = \Delta h + \Delta \delta \tag{3.7}$$

The "contact point" (zero ΔX) was taken to be the point at which the linear compliance line reached zero force, followed by the previous protocols on deformable surfaces [10, 11]. Before contact, ΔX represents the separation plus the relatively small deformation of the bubble which depends on the surface force between the probe and the bubble. After contact, ΔX represents only the deformation of the bubble because the separation between the probe and the bubble surface is considered to be zero. In the constant compliance region, the cantilever and the bubble are assumed as two springs in a series where the measured stiffness k_m is given by

$$\frac{1}{k_m} = \frac{1}{k_c} + \frac{1}{k_b} \tag{3.8}$$

Thus the bubble stiffness is given by

$$k_b = \frac{k_c}{\frac{k_c}{k_m} - 1} = \frac{k_c}{\frac{\alpha_{bubble}}{\alpha_{hard}} - 1} \tag{3.9}$$

where α_{hard} is the cantilever inverse optical lever sensitivity (InvOLS) against a hard surface and α_{bubble} is the cantilever inverse optical lever sensitivity (InvOLS) against the bubble.

The bubble stiffness can be also calculated by

$$k_b = \frac{F_b}{\delta} = \frac{F}{\delta} \tag{3.10}$$

since for two springs in series, $F = F_b = F_c = k_b \times \delta = k_c \times Z_c$.

Attard et al. described a theoretical way to express the stiffness of a bubble or droplet with

$$k_b = \frac{-4\pi\gamma}{\frac{\cos\theta}{2+\cos\theta} + \ln\left[\frac{R}{2\kappa R_b^2} \times \frac{(1+\cos\theta)^2}{\sin^2\theta}\right]} \tag{3.11}$$

which showed the bubble stiffness to be linearly dependent on the surface tension γ, and logarithmically depended on the decay length of the interaction κ^{-1}, the radius of the bubble R_b, the radius of the probe R, and the contact angle θ [12, 13].

A Teflon slide was cleaned in concentrated nitric acid for several minutes, followed by thorough rinsing with Milli-Q water. Air bubbles were spontaneously transferred from an Eppendorf pipette onto a spot on the Teflon slide where was immersed in Ludox nanoparticle suspensions. The bubble diameter was typically 800 μm as determined by top view light microscopy connected to AFM (Fig. 3.4). Gas bubbles are thermodynamically unstable and tend to dissolve in water due to the Laplace pressure [14]. However air bubbles are much more stable when existing in colloidal nanoparticle suspensions, probably because the particles prevent coalescence of bubbles [15]. The rest parts of the measurement followed the cases of on the solid substrates.

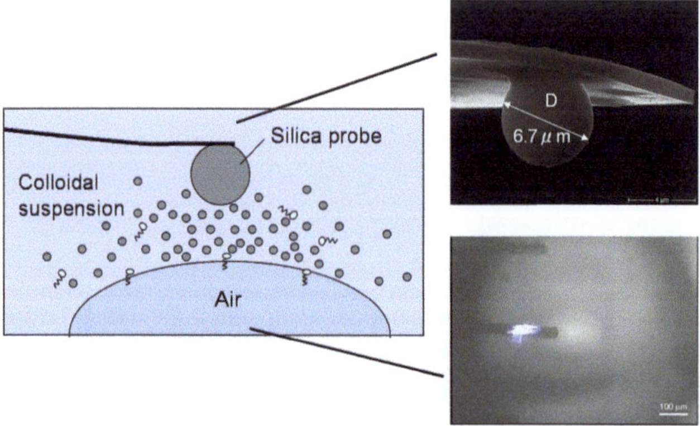

Fig. 3.4 *Left* schematic representation of the AFM setup for the force measurements. *Right top* scanning electron microscope image showing the silica microsphere glued to the end of the AFM cantilever. *Right bottom* view from the top showing placement of cantilever probe right on the top centre of the air bubble surface. The middle cantilever at which the laser beam aligned is in the focus. The brightest part of the ring underneath is the top centre of the bubble

To quantitatively study the structuring of nanoparticles, the oscillatory forces are fitted with Eq. 2.14 as well. Based on Derjaguin approximation, the bubble which is $800 \, \mu m$ in diameter can be considered as flat surface.

3.3.2 Small Angle X-ray Scattering

Small angle X-ray scattering (SAXS) is an accurate and non-destructive analytical method to determine the particle's structure in terms of particle size and shape. The particle sizes can be resolved in a range from 1 to 50 nm between the typical $0.1°$ and $10°$ of scattering angles. An electron density difference between particles and solvent is required to establish contrast in SAXS. Besides the interaction between the incoming radiation and particles, a detector is also needed to record and reconstruct scattering patterns of particles. In the recording process the phases of the detected waves are lost. Because the whole illuminated sample volume is investigated, the average values of the structure parameters are obtained with SAXS. The signal amplitude scales with the square of volume of the particle V_p, the particle number density ρ and the square of the contrast ΔSLD^2. The closer the lens is to the object (the larger the scattering angle), the smaller is the detail that can be resolved. Under Bragg relation is valid, the length scale probed in the experiment is related to measurable parameter q as

$$d = \frac{2\pi}{q} \tag{3.12}$$

where $q = \frac{4\pi}{\lambda} \sin \theta$, 2θ being the scattering angle.

The SAXS measurements were performed on a new version of small angle X-ray equipment-SAXSess (Anton Paar, Graz). The equipment consisted of a sealed tube to generate X-ray (Cu Kα, 0.1542 nm) and a line collimation system. Sample-detector distance was 309 mm. A fluid flow cell with a 1 mm quartz capillary was used. For each sample, the output intensity was the integral of 100 frames of measurements. Data treatment was done using SAXSquant 3.5 (Anton Paar, Austria). The data were first normalized using the primary beam intensity as a standard. The water background was subtracted and then the desmearing process was performed with used beam length and width profiles. At the end, the structure factor was extracted out by dividing the form factor from the total intensity. The structure peak has a Lorentzian line shape, produced from a Fourier transformation on a complex exponential function of g(r)$\propto$ e$^{-r/(\frac{2}{\Delta q})}$ cos($q_{max}r$) [16, 17]. The structure peak is fitted with

$$S(q) = \frac{S_0 \left(\frac{\Delta q}{2}\right)^2}{(q - q_{max})^2 + \left(\frac{\Delta q}{2}\right)^2} + y_0 \tag{3.13}$$

where S_0 denotes the structure peak intensity, Δq the full width at half maximum of the intensity, q_{max} the center, and y_0 the baseline of the peak.

3.3.3 Other Methods for Solution Characterization

3.3.3.1 Surface Tension Measurements

Surface tension is the cohesive energy present at an interface, describing the property of a liquid to resist external force. The interactions of a liquid molecule in the bulk are balanced by an equal attractive force with surrounding molecules in all directions. Molecules on the surface of a liquid experience an imbalance of forces, which is energetically unfavorable. In order to bring a molecule from bulk to the surface, extra work is needed. This work dW which is proportional to the number of molecules brought to the surface from the bulk and thus to the surface area dA can be presented as

$$dW = \gamma dA \tag{3.14}$$

The constant γ is the surface tension and has the dimension of energy per unit area $(\mathrm{J\,m^{-2}})$ or force per unit length $(\mathrm{N\,m^{-1}})$.

The surface tension of both the pure silica nanoparticle suspensions as well as mixtures of nanoparticle suspensions and various surfactants were measured via a K11 Tensiometer (Krüss, Germany) under clean room conditions. The du Noüy ring method [18] was used with a thin Ir-Pt wire ring with the radius R_{ring}. The surface tension is obtained from the force needed to balance the liquid meniscus before the ring is detached from the liquid surface [19]

$$\gamma = \frac{F}{4\pi R_{ring}} \tag{3.15}$$

The experiments were performed at $25\,^\circ\mathrm{C}$ in a Teflon vessel (diameter of $5\,\mathrm{cm}$). Before each measurements the vessel was equilibrated for at least $15\,\mathrm{min}$.

3.3.3.2 Zeta-Potential Measurements

Zeta potential is the electric potential difference between the stationary layer of fluid at the slipping plane in the diffuse double layer and the dispersion medium. Since the zeta potential indicates the degree of repulsion between adjacent likely charged particles, the magnitude of zeta potential can be related to the stability of colloidal suspensions. The higher the zeta potential, the higher the stability of the colloids. The zeta-potential was measured via a Zetasizer Nano ZS (Malvern Instruments, Germany). An electric field is applied across the suspension. Particles in the suspensions move toward the electrode of opposite charge. The frequency shift or phase shift of an incident laser beam caused by the moving particles is measured as the particle mobility, and this mobility is converted to the zeta potential using the dispersant viscosity η and dielectric permittivity ϵ in the Smoluchowski equation

$$U_E = \frac{\epsilon \zeta}{\eta} \tag{3.16}$$

3.3.3.3 Conductivity Measurements

A conductometer "WTW series inolab pH/cond" with a cell "TetraCon 325" was used. The cell constant is $0.475\,cm^{-1}$, thus the conductivity measurable ranges go from 0.5 to $2000\,\mu S\,cm^{-1}$. The conductivity of samples was measured at room temperature and converted to the ionic strength with individual prefactor for each sized nanoparticle suspensions.

3.3.4 Other Methods for Surface Characterization

3.3.4.1 Contact Angle Measurements

The contact angle is the angle between a liquid/vapor interface and a solid surface, which is a measure of the interaction across three phases. Based on the spreading behavior of a medium on a solid surface, the contact angle can be varied from 0° to 180° according to the hydrophobicity of the solid surfaces. The contact angle can be calculated by Young's equation [20] in the thermodynamic equilibrium status

$$\gamma_{LG} \cos\theta = \gamma_{SG} - \gamma_{SL} \tag{3.17}$$

where θ is the contact angle, γ_{LG}, γ_{SG} and γ_{SL} is the surface tension of liquid-vapor, solid-vapor and solid-liquid interface, respectively.

The contact angle of silica nanoparticle and surfactant mixture solution on silicon wafer was determined with dynamic sessile drop method by an OCA 20 device from Dataphysics (Germany) under ambient conditions. The liquid droplet profile was captured with optical subsystem and contact angle was assessed directly by measuring the angle formed between the baseline of the solid surface and the tangent to the drop contour by image analysis. Both static and dynamic measurements were able to be performed.

3.3.4.2 Ellipsometry

Ellipsometry has been widely used to determine the film thickness of mono- or multilayer coated on a substrate. Ellipsometry setup normally includes five parts, the light source, incident beam polarizer, the sample stage, analyzer for reflected beam off sample, and the detector. Ellipsometry measures the change of polarization upon reflection. This change is related to the sample thickness and dielectric properties. Measurements were performed with a Multiscope from Optrel GbRm (Wettstetten, Germany) in Null-Ellipsometry mode. A He-Ne laser with wavelength of 632.8 nm was used, the angle of incidence and reflection were set to be the same at 70°. Alignment was needed before each measurement to make sure the reflected beam was located in the center of the detector. The complex reflectance ratio between p-polarized (r_p) and s-polarized reflected beam (r_s) can be parametrized by the measured values of amplitude ratio Ψ and the phase shift Δ.

$$\tan(\Psi)e^{i\Delta} = \frac{r_p}{r_s} \tag{3.18}$$

The instrument was controlled by the software Multi, which measures Ψ and Δ. The data analysis for the determination of thickness d and refractive index n of the multilayer was performed by using the software Elli (Optrel). A model analysis with four layers was used; (i) air ($n = 1$), (ii) multilayer, (iii) SiO_x ($d = 1.5$ nm, $n = 1.4598$) and (iv) Si ($n = 3.8858$, $k = -0.02$) where k is the imaginary part of the refractive index which is related to the so-called extinction coefficient.

References

1. Decher, G. (1997). *Science, 277*, 1232–1237.
2. Sukhorukov, G., Donath, E., Davis, S., Lichtenfeld, H., Caruso, F., Popov, V., et al. (1998). *Polymers for Advanced Technologies, 9*, 759–767.
3. Binning, G., Rohrer, H., Gerber, C., & Weibel, E. (1982). *Physical Review Letters, 49*, 57–61.
4. Ducker, W., Senden, T., & Pashley, R. (1991). *Nature, 353*, 239–241.
5. Butt, H. (1991). *Biophysical Journal, 60*, 1438–1444.
6. Ducker, W., Senden, T., & Pashley, R. (1992). *Langmuir, 8*, 1831–1836.
7. Hutter, J., & Bechhoefer, J. (1993). *Review of Scientific Instruments, 64*, 1868–1873.
8. Dagastine, R., Stevens, G., Chan, D., & Grieser, F. (2004). *Journal of Colloid and Interface Science, 273*, 339–342.
9. Hoh, J., & Engel, A. (1993). *Langmuir, 9*, 3310–3312.
10. Fielden, M., Hayes, R., & Ralston, J. (1996). *Langmuir, 12*, 3721–3727.
11. Butt, H., Cappella, B., & Kappl, M. (2005). *Surface Science Reports, 59*, 1–152.
12. Attard, P., & Miklavcic, S. (2001). *Langmuir, 17*, 8217–8223.
13. Attard, P., & Miklavcic, S. (2003). *Langmuir, 19*, 2532.
14. Epstein, P., & Plesset, M. (1950). *Journal of Chemical Physics, 18*, 1505–1509.
15. Binks, B. (2002). *Current Opinion in Colloid and Interface Science, 7*, 21–41.
16. Helm, C., Moehwald, H., Kjaer, K., & Alsnielsen, J. (1987). *Europhysics Letters, 4*, 697–703.
17. Spaar, A., & Salditt, T. (2003). *Biophysical Journal, 85*, 1576–1584.
18. Noüy, L. J. D. (1919). *Journal of General Physiology, 1*, 521.
19. Schwuger, M. J. (1996). *Lehrbuch der Grenzflächenchemie*. Stuttgart, Germany: Georg Thieme Velag.
20. Young, T. (1805). *Philosophical Transactions of the Royal Society of London, 95*, 65.

Chapter 4
Structuring of Nanoparticle Suspensions Confined Between Two Smooth Solid Surfaces

4.1 Introduction

Confining particles between two solid surfaces leads to damped oscillatory forces [1, 2]. This well-known effect is directly related to the oscillating particle density profile perpendicular to the surface [3, 4]. The oscillatory force occurs when the oscillating concentration profile of the particles in front of the opposing confining surfaces overlap. With decreasing separation between the two confining surfaces, the layers of particles are pressed out one after another, which leads to measurable alternating repulsion and attraction. The oscillatory force thus indicates the periodic layering of confined particles. The force can stabilize the colloidal systems, since it hampers drainage of the film [5, 6]. The oscillatory wavelength represents the distance between two adjacent layers of particles formed parallel to the confining surfaces. The decay length is a measure of how far particles correlate to obtain periodic oscillations. There exists presently several techniques such as the surface force apparatus, [4] total internal reflection microscopy, [1] optical tweezers, [7] thin film pressure balance, [8, 9] and colloidal probe atomic force microscopy [10] to measure the oscillatory forces.

The first study of the ordering of colloidal particles can be traced back to the 1980s. Nikolov et al. found that thinning films of aqueous dispersions of polystyrene latex nanoparticles changed thickness with regular step-wise abrupt transitions by using reflected light microinterferometry [11]. These observations verified that the step-wise thinning or stratification of thin liquid films could be explained as a layer-by-layer thinning of ordered structuring of colloidal particles formed inside the film. There are several other papers that have also shown that particles tend to form periodic ordering during the approach of confining surfaces by methods of thin film pressure balance [12–14] and total reflectometry [15, 16].

Recently, the structuring formation has been studied by the measurement of the oscillatory force of colloidal particles by Piech and Drelich et al. [17–21] with colloidal probe atomic force microscopy (CP-AFM), which was advantageous in measuring the complete oscillatory force curves for various systems [17–27].

Y. Zeng, *Colloidal Dispersions Under Slit-Pore Confinement*, Springer Theses,
DOI: 10.1007/978-3-642-34991-1_4, © Springer-Verlag Berlin Heidelberg 2012

Among those mentioned studies, the oscillatory wavelength of colloidal particles was found to depend on the bulk particle volume fraction ϕ according to $\lambda \propto \phi^{-1/3}$ at relatively low volume fraction [17, 18, 20, 21]. At sufficiently high volume fraction, the wavelength was found to be close to the effective particle diameter, $2(R + \kappa^{-1})$ [19].

However, a precise understanding of the characteristic lengths, that is the wavelength and decay length (correlation length) of the oscillations in relation to the corresponding bulk properties and their dependence on the internal and external sample properties, is still missing [19, 20].

In this chapter, AFM, small angle X-ray scattering (SAXS), and theoretical modelings are combined to investigate the interaction in suspensions of charged silica nanoparticles and test the validity of the DFT predictions in a real colloidal fluid. The interparticle distance and correlation length in bulk as obtained from SAXS are compared to those found under confinement as obtained from AFM. Both experimental results are compared to the theoretical results in the framework of Derjaguin-Laudau-Verwey-Overbeek (DLVO) theory, where the interaction between two nanoparticles is described via a screened Coulomb potential. Three different-sized silica nanoparticles, with mean particle diameters of 11, 16 and 26 nm are used. The geometric confinement effect on the ordering of nanoparticles is studied by comparing the change of characteristic lengths. The dependence of each characteristic length on variation of particle size, particle concentration, and ionic strength, and their power-law are investigated. The interaction strength, force amplitude and maximum scattering intensity, in relation to the particle concentration and particle size, is discussed as well.

4.2 Results

4.2.1 Effect of Confinement and Particle Concentration

In order to know the effect of confinement, the structuring of silica nanoparticles in bulk was first determined by small angle X-ray scattering (SAXS).[1] Figure 4.1a shows the SAXS diagram of Ludox silica nanoparticle suspensions with particle diameter of 26 nm at varying particle concentration. With increasing sample concentration, the structure peak position q_{max} shifts to the high q region. The grey lines in the figure are the corresponding form factor $F(q)$ calculated using the polydisperse sphere model. It is apparent that the form factor does not change with concentration, thus the structure factor can be extracted by dividing the form factor $F(q)$ from the total intensity. Figure 4.1b shows the corresponding structure factor with fitting curve by Lorentzian equation (Eq. 3.13), from which the quantitative values of q_{max} and Δq can be obtained. As particle concentration increases, the position of maximum

---

[1] Reprinted with permission from: *Surviving Structure in Colloidal Suspensions Squeezed from 3D to 2D*, Sabine H. L. Klapp, Yan Zeng, Dan Qu, and Regine von Klitzing, *Physical Review Letters*, **2008**, *100*, 118303. Copyright (2008) by the American Physical Society.

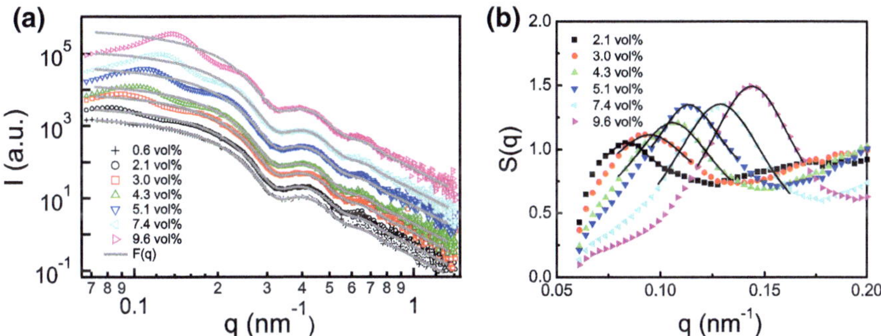

Fig. 4.1 **a** SAXS diagrams of Ludox nanoparticle suspensions with particle diameter of 26 nm at different concentrations. *Grey lines* represent the best form factor $F(q)$ fitted with polydisperse sphere model. **b** The structure factor extracted from SAXS intensity. Peaks fitted with the Lorentzian form of Eq. 3.13

q_{max} shifts to higher q values and its width Δq increases. Under the assumption that the Bragg relation is valid, the mean particle distance is the reciprocal of the peak position, $2\pi/q_{max}$, which decreases with increasing particle concentration. In addition, $2/\Delta q$ corresponds to the decay length of pair correlation function $g(r)$, thus it can be also called the correlation length of the particle interaction. In this section the investigation of the wavelength is focused, while in the next the correlation length is further discussed.

To which extent the bulk wavelength λ_b persists in the presence of confinement is now investigated. Experimental results for the oscillatory force $F(h)$ were recorded with CP-AFM, in which nanoparticles were confined between a silica micro-sphere glued on the AFM cantilever and a silicon wafer. Figure 4.2 shows some examples of AFM force versus distance curves at varying particle concentration. For all but the highest concentration considered, the data are well fitted after the first minimum ($h > h_{min}$) by an exponentially damped oscillation with wavelength λ_f based on Eq. 2.14. The fitting curves by Eq. 2.14 are shown in Fig. 4.2 as solid lines. Moreover, the data clearly show that λ_f decreases and the oscillations become more pronounced with increasing particle concentration. At the highest concentration (10.9 vol%) one observes a deviation from the fit function for the first maximum (of about $1\sigma - 2\sigma$), indicating a different spatial distribution in ultrathin films of the last few layers. This different distribution may be partially due to crystallization effects close to the surfaces.

The corresponding experimental results for λ_f as a function of the particle volume fraction are plotted in Fig. 4.3a. Also shown are the theoretical GCMC data for λ_f (which, as demonstrated in Fig. 2.4, equals the bulk wavelength λ_b plotted in Fig. 2.3), and the experimental SAXS data for the bulk wavelength $\lambda_s = 2\pi/q_{max}$ deduced from the structure factor. Clearly, there is good agreement between the experimental data for λ_f and λ_s. λ_s is considered as a wavelength averaged over all particle separations, as the structure factor $S(q)$ is the Fourier transform of the *full* function

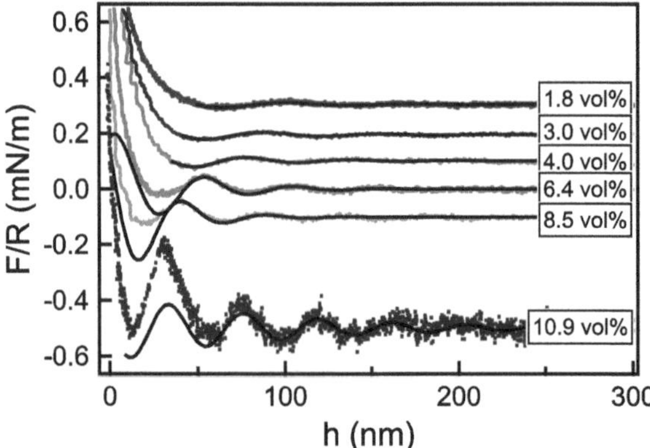

Fig. 4.2 Experimental curves for normalized force $F(h)/R$ obtained by CP-AFM for three different particle concentrations (the data have been vertically offset for ease of viewing). The curves are fitted according to the formula (2.14)

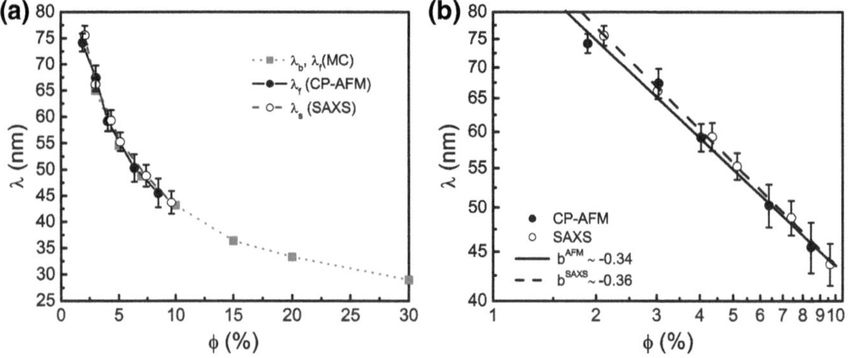

Fig. 4.3 **a** Comparison of the various wavelengths from theory and experiment in bulk and confinement. Not included are the theoretical results for λ_s since these are very similar to the SAXS data (see Fig. 2.3). **b** shows the experimental data on a logarithmic scale

$h_b(r)$ involving all poles, which does not need to coincide with theoretical bulk wavelength λ_b. The latter determines the asymptotic behavior via the leading pole (Eq. 2.19). Still, one expects these two wavelengths to be very close to each other. This is confirmed by the MC results for λ_b, which coincide well with the experimental data for λ_s (see Fig. 4.3a). Thus, the experimental data for λ_s is considered as an accurate approximation of the true wavelength λ_b characterizing $g_b(r)$ in the real bulk system. Therefore, the experimental AFM and SAXS results are completely consistent with the DFT prediction $\lambda_f = \lambda_b$ [28, 29]. Moreover, one see from Fig. 4.3 that there is excellent agreement between experimental and theoretical data for λ_f.

This is strong yet indirect evidence that the actual shape of the fluid-wall interactions (which is simplified in the theoretical model) is irrelevant for the asymptotic

Table 4.1 Experimental and theoretical results for the exponents b of the wavelengths resulting from a fit according to $\lambda = a\phi^{-b}$

Type	Experiments	Theories
Bulk	$b^{SAXS} = 0.36$	$b^{MC} = 0.36, b^{HNC} = 0.39$
Confinement	$b^{AFM} = 0.34$	$b^{GCMC} = 0.36$

decay of surface forces, which conforms with the DFT predictions [28, 29]. It is also noted that, irrespective of the concentration considered, the amplitudes and phases characterizing the experimental data are different from those of the theoretical functions $F(h)$ illustrated in Fig. 2.4. This is expected in view of the simplified fluid-wall potential $u_{fw}(z)$ used in the theoretical calculations (Eq. 2.20). The influence of fluid-wall potential on the amplitudes and phases of the oscillatory forces will be addressed in Chap. 5.

The very similar behavior of the wavelengths λ_f and λ_b is also reflected by the close values of the exponents b governing their power-law density dependence (i.e., $\lambda = a\phi^{-b}$ shown in Fig. 4.3b). The exponents are shown in Table 4.1. Indeed, for λ_b that $b^b = 0.36$ from experiment, while for λ_f, $b^f = 0.34$ are found. The theoretical results for the exponents are close to the experimental ones as well.

4.2.2 The Influence of Salt

The main goal in the following section is to identify the effect of the salt concentration, or I_{salt}, on the oscillatory force and the related structuring.[2] As a starting point, the results of SAXS experiments in bulk system are considered. The structural factors of particles at 5.1 vol% and varying salt concentrations are shown in Fig. 4.4, where the peak broadening and intensity decrease is observed, indicating the correlation length $\xi_b = 2/\Delta q$ decreases with increasing salt concentration. The mean particle distance $\lambda_b = 2\pi/q_{max}$ remains the same up until the point that particles start to form aggregates at concentrations higher than 10^{-3} M (see Table 4.2). These are consistent with the results for theoretical modelings, [30] where a decrease of the correlation length ξ_b and constant wavelength λ_b with increasing salt concentration have been observed.

The experimental results for $F(h)$ from CP-AFM measurements at volume fraction of $\phi = 7$ vol% and five different salt concentrations obtained by adding none, 10^{-5}, 10^{-4}, 10^{-3}, and 10^{-2} M of NaCl to the system are shown in Fig. 4.5. The force amplitude decreases significantly with increasing salt concentration. Moreover, at a NaCl

---

[2] Reprinted with permission from: *Asymptotic structure of charged colloids between two and three dimensions: the influence of salt*, Sabine H. L. Klapp, Stefan Grandner, Yan Zeng, and Regine von Klitzing, *Journal of Physics: Condensed Matter*, **2008**, *20*, 494232. Copyright (2008) by the Institute of Physics.

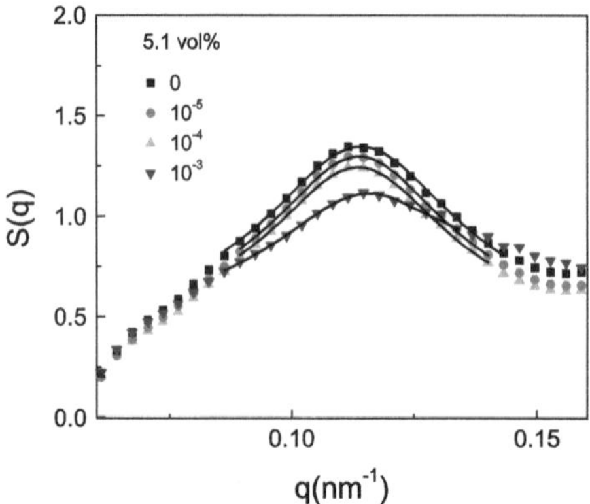

Fig. 4.4 The structure factor of 26 nm sized particle suspension with varying salt concentration at a fixed particle concentration. *Solid lines* are the fits according to Lorentzian Eq. 3.13

Table 4.2 SAXS results for the mean particle distance $\lambda_b = 2\pi/q_{max}$ and the correlation length $\xi_b = 2/\Delta q$ at different concentration of added NaCl and a fixed particle concentration of 5.1 vol%

I_{salt} (M)	$2/\Delta q$ (nm)	$2\pi/q_{max}$ (nm)
0	35.7	55.2
10^{-5}	34.8	55.3
10^{-4}	32.7	55.4
10^{-3}	26.5	55.1

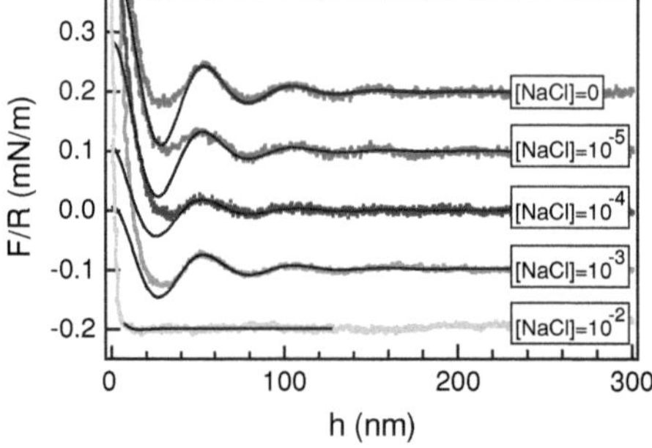

Fig. 4.5 Experimental curves *F(h)/R* obtained by CP-AFM at five different salt concentrations and a given particle volume fraction $\phi = 7$ vol% (the data have been vertically offset for clarity). The salt concentrations were adjusted by addition of NaCl as indicated (I_{salt} is given in M). The *solid lines* are fits according to Eq. 2.14

concentration of 10^{-2} M the oscillations of the force have essentially disappeared. This is consistent with GCMC simulations, [30] where a primary effect of adding salt consists of a pronounced decrease in the amplitude of the oscillations and the oscillations essentially vanish at $I_{salt} \geq 10^{-3}$ M. Similar results were obtained for confined polyelectrolytes solutions where the oscillations were drastically reduced after adding salt well below the ionic strength induced by the polyelectrolytes [31].

The solid lines in Fig. 4.5 are the fitting curves obtained according to Eq. 2.14. The fit describes the experimental curves at distances larger than the first minimum quite well, but not on shorter length scales. According to Eqs. 2.19 and 2.21, the expression Eq. 2.14 describes only the asymptotic behavior of the oscillatory force. That means, the breakdown of Eq. 2.14 at small h is rather expected. In addition, the deviation from the fit at shorter lengths could also be due to the relatively low spring constant of the cantilever used in the measurement compared to the strong attractive force caused by the exclusion of the layers of particles, which leads to mechanical instability in those regions of the force curves. Interestingly, the deviation from the asymptotic behavior at small h becomes less pronounced with increasing salt concentration. This behavior indicates that the increased electrostatic screening within the system lowers the surface forces and results in reduced mechanically instability. At higher particle concentration (e.g. 10.9 vol% in Fig. 4.2), the increasing of the electrostatic screening can lower the tendency for ordering and/or crystallization next to the surface as well. In the case of absence of crystallization, the optimal fitting needs to cover the first peak of oscillation instead of the valley for the aforementioned reason.

The influence of the salt concentration on the wavelength λ is now considered in more detail. The experimental SAXS results for λ_b and HNC results of λ_b obtained from a pole analysis of the corresponding bulk correlation function are included in Fig. 4.6a. The data from MC approach have the same values as in Fig. 4.6b and they are rarely influenced by salt concentration, therefore only one representative plot at $I_{salt} = 10^{-5}$ M is shown. Experimental CP-AFM results for λ_f as a function of the particle volume fraction and four salt concentrations are plotted in Fig. 4.6b. Also shown are (GC)MC data for λ_f which, as explained above, is equal to the

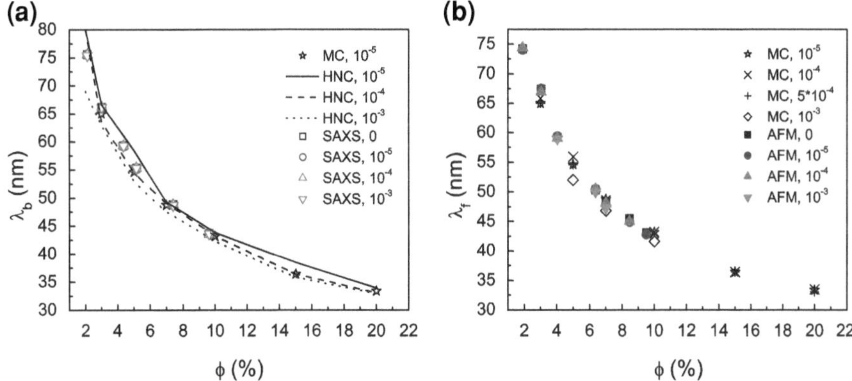

Fig. 4.6 Asymptotic wavelength as a function of the volume fraction at different salt concentrations I_{salt} (in M). **a** MC and HNC results for the bulk system and experimental data from SAXS measurements. **b** (GC)MC simulation results (where $\lambda_f = \lambda_b$) and experimental data from CP-AFM measurements

leading *bulk* wavelength λ_b characterizing the MC bulk correlation functions. This is consistent with the DFT predictions [29] and also with the previous findings of particles at zero I_{salt} in Sect. 4.2.1. Each of the approaches (CP-AFM, MC, HNC) consistently predicts that variation of the salt concentration has only a very small effect on the actual value of λ, the differences between I_{salt} from 10^{-5} to 10^{-3} M are essentially negligible. Moreover, Fig. 4.6 shows that there is excellent agreement between SAXS and HNC/MC data for λ_b and AFM and (GC)MC simulation data for λ_f.

In addition, all approaches yield close results indicate $\lambda_f = \lambda_b$ and predict a power-law behavior of the wavelength according to $\lambda = a\phi^{-b}$ with b $\approx$ 1/3, corresponding to an isotropic structuring. Thus, although the system is confined and characterized by layer formation (i.e. translational symmetry is broken), the average interparticle distance along the direction normal to the surface remains the same as that in the isotropic bulk phase.

Followed, the asymptotic correlation (decay) lengths, ξ, both in confinement and in bulk are addressed. ξ is a measure of the range over which particles in one region are correlated with those in another region. A smaller ξ indicates a smaller interaction distance, which corresponds to a less ordered structure. Figure 4.7 shows the comparison of ξ_f, of AFM force curves have been fitted with Eq. 2.14, and $\xi_b = 2/\Delta q$ of SAXS structure factors have been fitted with Lorentzian equation (Eq. 3.13). An excellent agreement between experimental ξ_f and ξ_b is shown. The fact that the oscillation decay length is equal to the range of positional correlations extracted from the SAXS peak width, suggests that the force decay is indeed mainly caused by the loss of positional correlations.

Regarding the theoretical results, the fact that the MC values λ_f in confined geometry are equal to those in the bulk system λ_b has been found for the wavelength. Having this in mind, the HNC and GCMC results for ξ_b and ξ_f, respectively, as a function of the volume fraction at different I_{salt}, in addition to experimental ξ_b and ξ_f are included in Fig. 4.8. Figure 4.8a shows the comparison of ξ_b between HNC and SAXS, while Fig. 4.8b shows the comparison of ξ_f between GCMC and

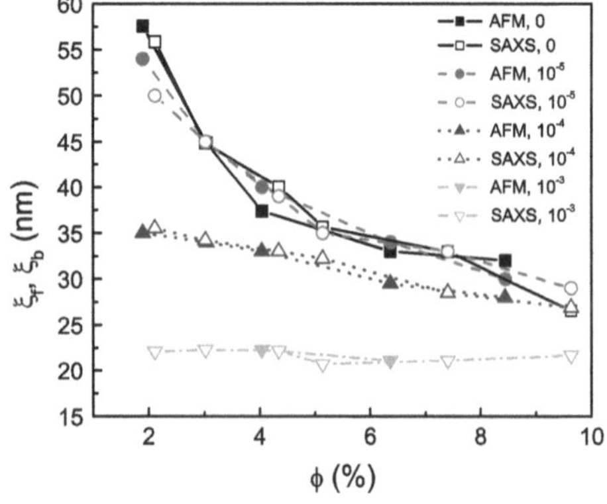

Fig. 4.7 AFM and SAXS results for the confined and bulk correlation length as a function of the volume fraction for different salt concentrations I_{salt} (in M). An excellent agreement is found, $\xi_f = \xi_b$

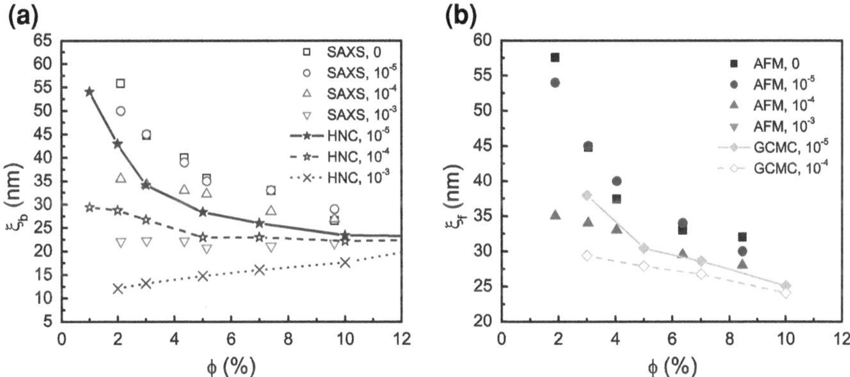

Fig. 4.8 **a** HNC results for the correlation length ξ_b as a function of the volume fraction at different salt concentrations I_{salt} (in M) are shown beside the experimental ones from SAXS. **b** The theoretical ξ_f obtained from GCMC in comparison to the experimental results from AFM measurements. The experimentally determined correlation lengths are larger than the theoretical ones

AFM. All approaches predict a significant influence of the salt concentration on the correlation length as long as the volume fraction is not too large, that is, $\phi \leq 10$ vol%. For smaller volume fractions, adding salt at a fixed silica concentration yields a pronounced decrease of ξ. This can be explained by simple screening arguments. For $I \geq 10^{-4}$ M, ξ increases monotonically with ϕ in the range of volume fractions considered. These strongly screened systems behave more like systems with hard repulsive potentials (i.e. hard spheres) where the range of oscillatory correlations (and thus, ξ) just increases with ϕ. On the other hand, for $I_{salt} \leq 10^{-4}$ M, a decrease of ξ is observed. This behavior may be interpreted as follows: for small values of I_{salt} and ϕ, increasing the silica concentration has a similar effect to adding salt since both yield an increase of the inverse Debye length κ and thus the screening. This leads, in turn, to a damping of oscillations and the related surface properties. It is worth to mention that at larger ϕ does the system with low salt behave again like a "hard-sphere", in that ξ increases with ϕ. These trends of the correlation length are predicted both by HNC theory and the MC data at $\phi > 10$ vol% [30].

At the particle concentrations considered, the experimentally determined decay (correlation) lengths are larger than the theoretical correlation lengths plotted in Fig. 4.8. This is also visible in the slower decay of the experimental force curves (see Fig. 4.5) in comparison to the simulated ones [30]. According to DFT, the different fluid-wall potential in experiments and theory should not have any effect on ξ, and ξ_f should be equal to ξ_b as long as the asymptotic limit is considered. This has been found experimentally by comparing results of AFM and SAXS (Fig. 4.7), but not between experimental and theoretical results.

Although HNC directly calculated the correlation functions from integral equations it contained several approximations which might cause the differences to the experimental results. Particularly in the theoretical GCMC data, there is clearly some uncertainty regarding the separation h where the asymptotic behavior actually sets in, the determination of the decay length ξ_f in Eq. 2.21 suffered from big uncertainties, whereas the wavelength λ_f yielded good agreement with the experimental results (see Fig. 4.6). The fitting procedure of the correlation function using Eq. 2.21 is only valid for the asymptotic range, i.e. for large distances. Choosing a fit range starting after several oscillations should yield better agreement between simulation and theoretical calculation. However, due to statistical errors of the simulations, the fitting procedure yielded erroneous results at large h since the amplitudes of $\tilde{f}(h)$ became very small at the bulk concentrations considered. Hence, the limited distance range to fit $\tilde{f}(h)$ yielded uncertainties in ξ. The same uncertainty was be found using Eq. 2.19 for the MC correlation length ξ_b. However, since HNC results yield better agreement with the experimental ones, MC data for bulk correlation length ξ_b is not included in Fig. 4.8a.

Nevertheless, the qualitative behavior of ξ, depending on the salt ionic strength I_{salt} and the particle concentration, was not affected by the fitting range. Further investigations of the correlation length and its other dependence will be discussed in the next section.

4.2.3 Impact of Particle Size

The main goal in this section is to identify the effect of particle size (and the resulting total particle surface charge $Z \propto \sigma^2$) on the structural forces in slit-pore confinement.[3] Three types of colloidal suspensions, named TMA 34, HS 40, and SM 30, composed of charged silica nanoparticles with different diameters σ were used. The particle sizes were determined by scanning electron microscopy (SEM) and by small angle X-ray scattering (SAXS).[4] The ζ-potential was determined by electrokinetic measurements at the same conditions employed in the AFM experiments. The corresponding particle diameters, zeta potentials, and total surface charge Z (see Sect. 2.2.1) are summarized in Table 4.3.

The scanning electron microscopy images are shown in Fig. 4.9. The images show that all three types of particles are highly spherical and characterized by a relatively small size distribution. A highly mono-dispersed system is necessary in order to determine the effect of particle size precisely, because the wavelength of a polydisperse system results from all contributions of particles with various sizes [17].

---

[3] Reproduced by permission of The Royal Society of Chemistry: *Effect of particle size and Debye length on order parameters of colloidal silica suspensions under confinement*, Yan Zeng, Stefan Grandner, Cristiano L.P. Oliveira, Andreas F. Thuenemann, Oskar Paris, Jan S. Pedersen, Sabine H. L. Klapp, and Regine von Klitzing, *Soft Matter*, **2011**, *7*, 10899–10909.

[4] Reproduced by permission of The Royal Society of Chemistry: *Charged silica suspensions as model materials for liquids in confined geometries*, Sabine H. L. Klapp, Stefan Grandner, Yan Zeng, and Regine von Klitzing, *Soft Matter*, **2010**, *6*, 2330–2336.

Table 4.3 Experimental results for the particle diameters σ, ζ-potentials, and total surface charge Z of the Ludox particles investigated in the study

Type	σ_{SEM} (nm)	σ_{SAXS}	ζ (mV)	Z
TMA 34	25 ± 2	26 ± 3	-60	35
HS 40	16 ± 2	16 ± 2	-57	13
SM 30	9 ± 2	11 ± 2	-56	6

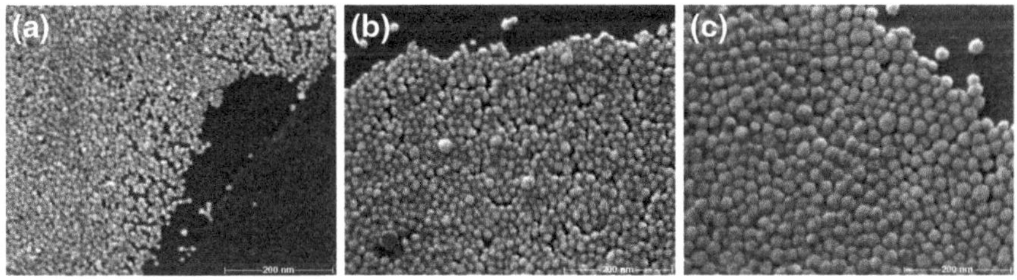

Fig. 4.9 SEM images of (**a**) SM 30 ($\sigma = 9 \pm 2\,$nm), (**b**) HS 40 ($\sigma = 16 \pm 2\,$nm), and (**c**) TMA 34 ($\sigma = 25 \pm 2\,$nm)

SAXS diagrams are shown in Fig. 4.10. Through fitting the form factor, particle size can be obtained, shown in Table 4.3. Furthermore one can observe, with increasing sample concentration, the structure peak position shifts to the high q region. The structure factor was extracted by dividing the form factor $F(q)$ from the total intensity. The corresponding structure factor for all three series of Ludox samples is shown in Fig. 4.11, with fitting curve by Lorentzian equation (Eq. 3.13), from which the quantitative values of q_{max} and Δq can be obtained. As particle concentration increases, or particle size decreases at a given particle concentration, the position of maximum q_{max} shifts to higher q values and its width Δq increases. The correlation length $2/\Delta q$ is reminiscent of the decay length of the oscillatory force measured by AFM under slit-pore confinement. The mean particle distance is the reciprocal of the peak position, $2\pi/q_{max}$, and can be compared to the wavelength of the oscillation from AFM measurements.

Figure 4.12 shows some examples of AFM force versus distance curves for Ludox silica nanoparticle suspensions with particle diameters of 11, 16 and 26 nm, respectively, at varying particle concentration. All curves in Fig. 4.12 exhibit oscillations indicating layer formation of the particles. In general, as described in previous sections, for the samples at a given particle size at higher concentrations, the force amplitude was more pronounced and the force range was larger, indicating a stronger interaction and more layers of particles. The wavelength of oscillation decreased with particle concentration in the meantime, which indicated that the layer-to-layer distance became smaller. The most prominent effect of varying particle size consists of a decrease in the wavelength λ_f of the oscillations upon decreasing σ. This change in λ_f confirms the idea that the particle diameter is an important length scale in the problem. A decrease in the amplitude of the oscillations, which became particularly

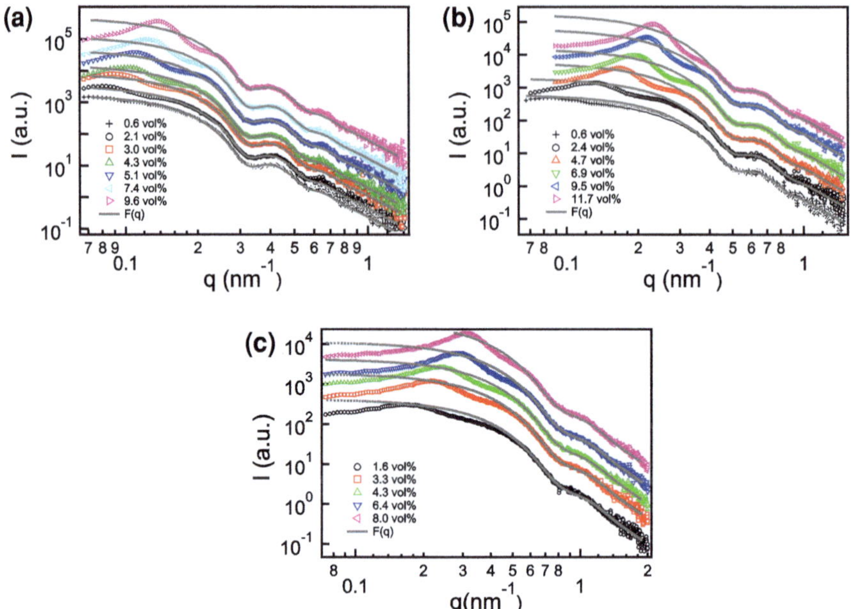

Fig. 4.10 SAXS diagrams of Ludox nanoparticle suspensions of three different particle sizes, (**a**) 26 nm (**b**) 16 nm and (**c**) 11 nm, at varying particle volume fractions. *Grey lines* represent the best form factor *F(q)* fitted with polydisperse sphere model

apparent for the smaller particles, is also observed. To quantify these effects and obtain quantitative values of λ_f, ξ_f and A, the curves (see solid lines in Fig. 4.12) are fitted according to Eq. 2.14.

The experimental results for λ_f at different volume fractions are summarized in Fig. 4.13, where includes the GCMC simulated results of theoretical λ_f for comparison. In GCMC simulations, negatively charged particles with $\sigma = 26$ nm and 16 nm are considered. $Z = 35$ is set for the larger particles and the total charge of the smaller particles with $\sigma = 16$ nm is set to $Z = 13$. The smallest particles with $\sigma = 11$ nm (corresponding to $Z \approx 6$) are not taken into account in the simulations, since the resulting DLVO repulsion is too small to generate detectable force oscillations at the volume fractions of interest. All GCMC simulations have been carried out with two different salt concentrations ($I_{salt} = 10^{-5}$ M and $I_{salt} = 10^{-4}$ M). The data in Fig. 4.13 reveal that the absolute value of I_{salt} is rather unimportant. On the other hand, there is a strong impact of the particle size on the wavelength of the oscillations, in agreement with the experimental results. This quantitative consistency underlines the validity of the DLVO-based model not only for TMA-34 particles ($\sigma = 26$ nm, $Z = 35$), which were studied earlier, but also for other particles sizes.

The exponents extracted from GCMC data are given in Table 4.4. While the values found at $\sigma = 26$ nm are still rather close to the experimental result (see Table 4.5), the GCMC values for $\sigma = 16$ nm deviate more significantly. However, a precise

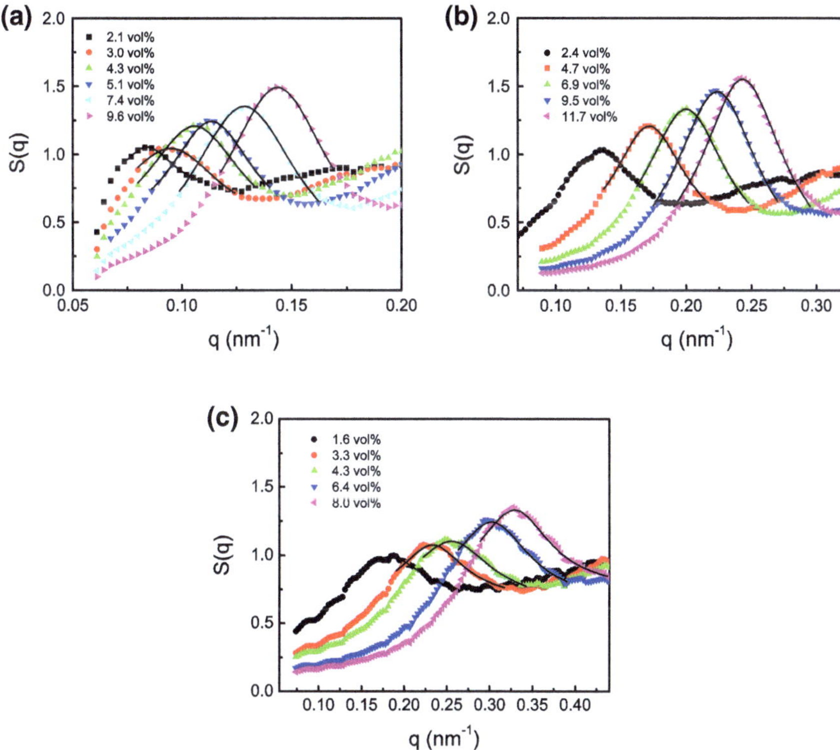

Fig. 4.11 The extracted structure factors of (**a**) 26 nm (**b**) 16 nm and (**c**) 11 nm sized particle suspensions. Peaks are fitted to the Lorentzian form of Eq. 3.13 to obtain the quantitative values of q_{max} and Δq

determination of λ_f at this small particle size is rather difficult, since the oscillations in $f(h)$ decay extremely fast [32]. Thus, it is hard to identify the "asymptotic" range relevant for which Eq. 2.21 should hold. Nevertheless, even with these slight deviations, one can conclude that the GCMC data confirms the idea of a bulk-like scaling ($b \approx 1/3$) of the wavelength within the range of volume fractions considered.

Even without a more detailed analysis one observes from the structure factors an increase of the peak width in Fig. 4.11 and from the force curves in Fig. 4.12 an increase of the damping of oscillations towards its limiting value of zero, that is, in another word a decrease of the correlation length ξ with increasing particle size. As expected from the relation between σ and Z, the larger (and thus, more strongly coupled) particles are characterized by more pronounced interparticle correlations. Since a quantitative measure of the range of correlations is the correlation length, the larger particles are characterized by larger correlation lengths. This is consistent with the theoretical results for ξ, [32] where MC results for the pair correlation functions $g(r)$ of two bulk silica suspensions composed of particles with $\sigma = 26$ nm and 16 nm are included. For MC data, the correlation lengths first decrease with

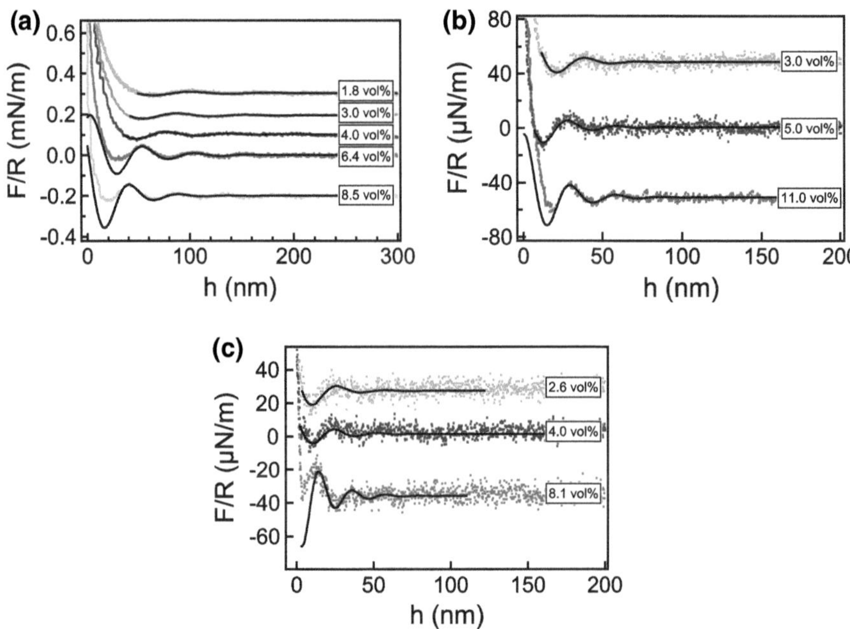

Fig. 4.12 AFM force curves of three series of Ludox nanoparticle suspensions, (a) 26 nm (b) 16 nm and (c) 11 nm, under slit-pore confinement. For better viewing, the curves are offset vertically. *Solid lines* are the corresponding curves fitted by Eq. 2.14

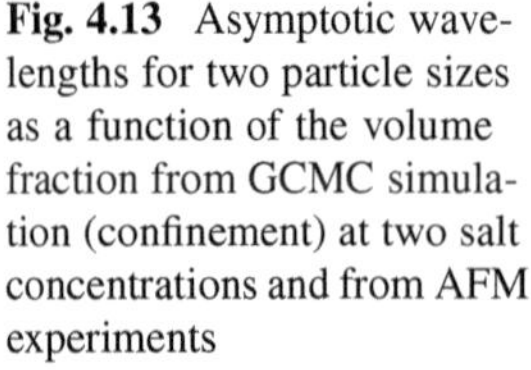

Fig. 4.13 Asymptotic wavelengths for two particle sizes as a function of the volume fraction from GCMC simulation (confinement) at two salt concentrations and from AFM experiments

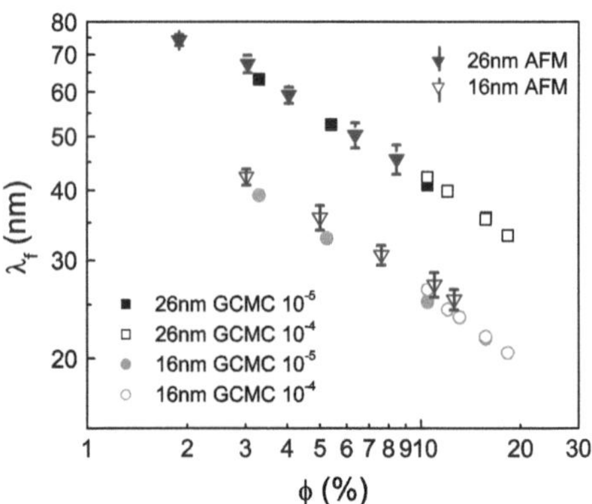

particle volume fraction till certain point and then increase again (hard repulsive potentials as aforementioned). In addition, similar correlation effects are observed in the microscopic structure of the confined systems: an increase of the size (and resulting charge) leads to both a stronger structuring in the z-direction, and more pronounced lateral correlations.

Table 4.4 GCMC results for the exponents b of the wavelengths resulting from a fit according to $\lambda = a\phi^{-b}$

Diameter (nm)	I_{salt} (M)	b
26	10^{-4}	0.38
26	10^{-5}	0.37
16	10^{-4}	0.46
16	10^{-5}	0.43

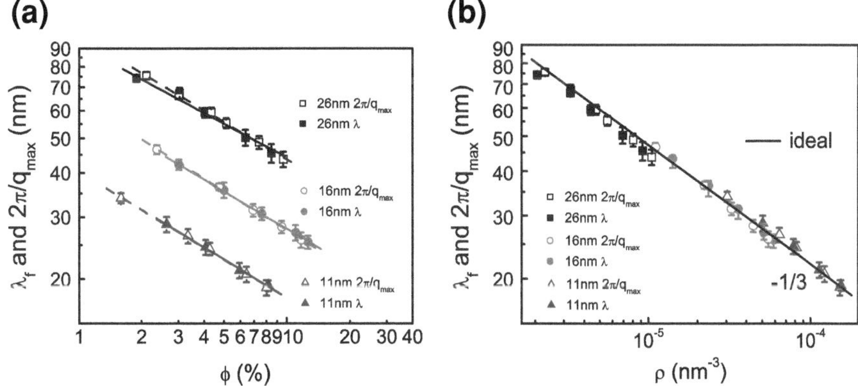

Fig. 4.14 a Comparison between AFM wavelength λ_f and SAXS $2\pi/q_{max}$ for three series of particles at varying concentrations. *Solid lines* are the corresponding fitted curves with scaling factor of approximately -0.33. **b** The master curve of Fig. 4.14a. The *solid line* is the calculated ideal value of average particle distance in bulk with $\lambda^{id} = \rho^{-1/3}$

4.3 Discussion

4.3.1 Scaling Law of the Interparticle Distance

Figure 4.14a shows the comparison of λ_f and $\lambda_b = 2\pi/q_{max}$ in a double logarithmic scale.[5] For all three series of Ludox samples, λ_f and $2\pi/q_{max}$ decrease with particle concentration and both values are in remarkable agreement. More precisely, as suggested by the location of the data points in Fig. 4.14a, the functions $\lambda(\phi)$ can be fitted according to a power law, i.e., $\lambda = a\phi^{-b}$. The resulting exponents are given in Table 4.5. One finds that all of the confined, layered systems essentially obey the simple (geometrical) *bulk* scaling rule according to which the wavelength should behave as the *mean* particle distance in an ideal (random fluid-like) system, i.e., $\lambda^{id} = \phi^{-1/3}$. One can also note that, the data sets for the three particle sizes in Fig. 4.14a are separated in the sense that the corresponding lines are shifted along the y-axis, i.e., the prefactors a in the power law $\lambda = a\phi^{-b}$ are size-dependent. This

[5] Reproduced by permission of The Royal Society of Chemistry: *Effect of particle size and Debye length on order parameters of colloidal silica suspensions under confinement*, Yan Zeng, Stefan Grandner, Cristiano L.P. Oliveira, Andreas F. Thuenemann, Oskar Paris, Jan S. Pedersen, Sabine H. L. Klapp, and Regine von Klitzing, *Soft Matter*, **2011**, *7*, 10899–10909.

Table 4.5 Experimental results for the exponents b of the wavelengths resulting from a fit according to $\lambda = a\phi^{-b}$

Type	b^{AFM}	b^{SAXS}
SM 30 ($\sigma \approx 11\,\text{nm}$)	0.33	0.33
HS 40 ($\sigma \approx 16\,\text{nm}$)	0.32	0.31
TMA 34 ($\sigma \approx 26\,\text{nm}$)	0.34	0.36

reflects the fact (already apparent from Fig. 4.12) that, at fixed volume fraction ϕ, the *absolute* value of λ decreases with σ. This is due to the different particle number densities at a given volume fraction for particles with unequal size, i.e. using smaller particles lead to a larger number density at the same volume fraction.

Given the rather simple behavior of the functions $\lambda(\phi)$, one may ask whether there is any *non-trivial* impact of the particle size on the wavelength. In other words, could one just "map" the data points for different diameters σ onto each other? To explore this question, the wavelength λ as a function of the particle number density, $\rho = N/V = (6/\pi)\phi\sigma^{-3}$, is plotted in Fig. 4.14b. In an ideal (random fluid-like) system one would expect that $\lambda^{\text{id}} = a\rho^{-1/3}$ with $a = 1$ (consistent with the scaling rule $\lambda^{\text{id}} = \phi^{-1/3}$ mentioned above). Given that this ideal scaling holds for the silica systems at hand, all data points should fall on one "master curve". In a double-logarithmic representation, this "master curve" is a line with slope $-b = -1/3$ and an intercept of $a = 1$. From the experimental data plotted in Fig. 4.14b one observe that the ideal scaling (indicated by the solid line) is fulfilled by all series of particles with an intercept of unity. The agreement of both wavelengths λ to the particle-size-independent ideal value indicates that the interparticle distance is solely number density dependent, not influenced by the particle size, whether in bulk or confinement.

It is worth mentioning, for samples prepared from different original stocks of suspensions, that when one uses the particle size determined from SEM, only the system TMA 34 composed of the largest particles has an intercept of approximately unity ($a = 1.04$). For the other systems, $a = 0.93$ for HS 40 and $a = 1.23$ for SM 30 (shown in Fig. 4.15a). The experimentally observed deviations from the ideal behavior in this context mainly stem from the uncertainty of the value of σ (see Table 4.3). This view is confirmed by Fig. 4.15b, where the same data with assuming somewhat different diameters is plotted (see figure caption). The fitted values for the intercept are now close to unity for all three systems considered ($a = 0.9988$, 0.9936, and 1.0095 for TMA 34, HS 40, and SM 30, respectively). Thus, a nearly ideal behavior is regained by adjusting particle sizes to 26 nm for TMA 34, 15 nm for HS 40, and 11 nm for SM 30, which approach the values determined from SAXS measurements. 1 nm smaller than 16 nm determined from SAXS for HS 40 is most likely due to the different stock of suspensions were used in the two measurements. Thus the size determined from SAXS is more accurate to be used as the mean particle sizes.

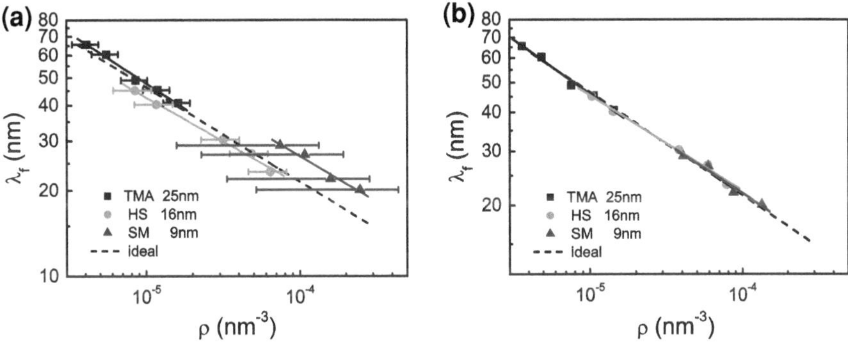

Fig. 4.15 **a** Wavelength as a function of the particle number density from CP-AFM experiments, assuming the average particle diameters from SEM given in Table 4.3. The *error bars* stem from the uncertainty in the particle diameters. The *dash line* corresponds to the ideal scaling rule $\lambda^{id} = \rho^{-1/3}$. **b** Same experimental data as in (**a**), with assuming somewhat different diameters (see figure caption)

4.3.2 Validity of $\lambda_f = 2(R + \kappa^{-1})$

Several literature studies have shown that the wavelength (in AFM, or step size in thin film pressure balance) coincided with the effective particle diameter $2(R + \kappa^{-1})$, including the Debye length, for colloidal samples at high concentration (above 10 vol%) [11, 19, 33]. The AFM wavelength λ_f with the calculated effective particle diameter, $2(R + \kappa^{-1})$, is herein compared. The contribution of the charge dissociation from the silica nanoparticle surfaces is needed to be taken into account on the total ionic strength I_{tot} and thus the Debye length κ^{-1}. There are two methods to determine or calculate the Debye length, calculate the κ^{-1} from Eq. 2.18 with known value of the silica surface potential (ζ-potential, see Table 4.3) or convert from conductivity of the suspension. Previous literature studies [20] used a simple Russell prefactor, 1.6×10^{-5}, which is valid for simple electrolytes, for conversion and yielded smaller values of the Debye length. Here, the prefactor for the present system is determined individually, based on the assumption of monovalent counterions and charge neutrality between counterions and colloidal particles. A relation between the measured conductivity and ionic strength of the samples is found to be $I_{tot} = \frac{|Z|}{2N_A\beta} \times K$, where β is the slope of the plot of conductivity versus particle number density as shown in Fig. 4.16. This equation yields a prefactor 1.27×10^{-6} to convert conductivity K in the unit of μS cm^{-1} to ionic strength I_{tot} in the unit of M for 26 nm sized particles. The values of Debye length κ_1^{-1} obtained from conductivity measurement are listed in Table 4.6, where the calculated values κ_2^{-1} from Eq. 2.18 are also included. The κ_2^{-1} and κ_1^{-1} have similar values, which are in the range approximately of $0.7 - 1.5\sigma$ for the particle concentration of interest.

It is obvious that the AFM wavelength λ_f is significantly smaller than the corresponding effective particle diameter in the absence of interactions between

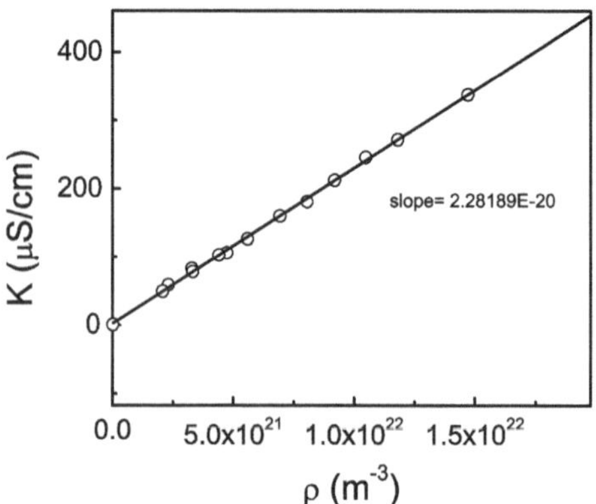

Fig. 4.16 Conductivity K versus particle number density ρ for 26 nm sized particles. The slope β is 2.282×10^{-20}

Table 4.6 Experimental results of conductivity K at varying particle concentration, the corresponding ionic strength and Debye length κ_1^{-1}, and the previous calculated Debye length κ_2^{-1} from Eq. 2.18, and the ratio of Debye lengths from two methods

ϕ (vol%)	K (μS cm^{-1})	I_{tot} (M)	mea. κ_1^{-1} (m)	cal. κ_2^{-1} (m)	$\kappa_2^{-1}/\kappa_1^{-1}$
2.1	58.8	7.4676×10^{-5}	3.519×10^{-8}	3.725×10^{-8}	1.058
3.0	82.8	1.0516×10^{-4}	2.966×10^{-8}	3.113×10^{-8}	1.049
4.3	106.1	1.3475×10^{-4}	2.620×10^{-8}	2.596×10^{-8}	0.990
5.1	126.5	1.6065×10^{-4}	2.399×10^{-8}	2.387×10^{-8}	0.995
7.4	181	2.2987×10^{-4}	2.006×10^{-8}	1.989×10^{-8}	0.992
9.6	246	3.1242×10^{-4}	1.721×10^{-8}	1.744×10^{-8}	1.014
1.8	49.3	6.2687×10^{-5}	3.841×10^{-8}	3.933×10^{-8}	1.024
3.0	77.9	9.8933×10^{-5}	3.057×10^{-8}	3.102×10^{-8}	1.014
4.0	102.6	1.3036×10^{-4}	2.663×10^{-8}	2.692×10^{-8}	1.010
6.4	160.2	2.0353×10^{-4}	2.131×10^{-8}	2.145×10^{-8}	1.006
8.5	212.1	2.6931×10^{-4}	1.853×10^{-8}	1.861×10^{-8}	1.004
10.9	271.6	3.4489×10^{-4}	1.637×10^{-8}	1.642×10^{-8}	1.003
13.5	337.8	4.2907×10^{-4}	1.468×10^{-8}	1.471×10^{-8}	1.002

particles, as shown in Fig. 4.17. This indicates that the particles' counterion double layers overlapped significantly in the present studied concentration range, leading to the strong electrostatic repulsion. In the meanwhile, the AFM results showed the $\rho^{-1/3}$ scaling law was valid at least until 13 vol% (the maximum experimentally studied concentration) and GCMC results extended the validity until 30 vol% [34]. Upon fitting the literature results which claimed to be close to the effective particle size, the scaling law of $-1/3$ was still obtained [11, 33]. Therefore, one can claim that interparticle distance is not ionic strength controlled and $-1/3$ scaling law is a general description for the distance of charged particles in the direction normal to the confining walls, as long as the repulsive interaction is sufficiently long-ranged. It is worth to mention that this scaling law is no longer valid when the interaction is characterized by the hard core of the particle, where the wavelength

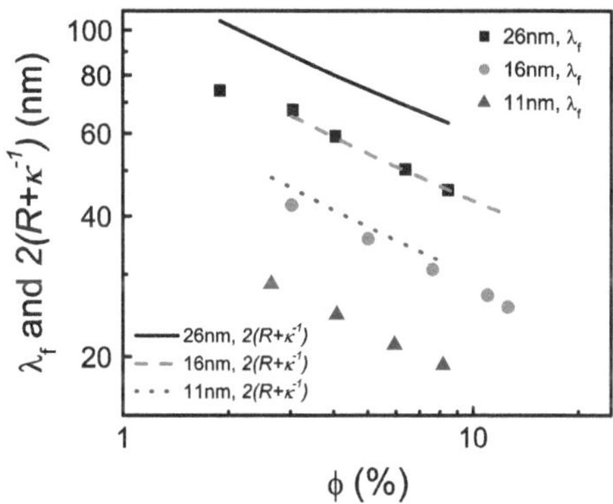

Fig. 4.17 Comparison between AFM wavelength λ_f and the calculated effective particle diameter, $2(R + \kappa^{-1})$. *Solid line*: $2(R + \kappa^{-1})$ for 26 nm, *dash line*: $2(R + \kappa^{-1})$ for 16 nm, *dot line*: $2(R + \kappa^{-1})$ for 11 nm

is the diameter of particle and not affected by the bulk concentration (discuss in detail in Chap. 7). The previous description of $2(R + \kappa^{-1})$ only in some systems (depends on ionic strength of the samples) approaches the value of wavelength at high concentrations.

4.3.3 Scaling Law of the Correlation Length

The comparison between ξ_f and $2/\Delta q$ for all three series of samples is shown in Fig. 4.18. There is a good agreement between ξ_f and $2/\Delta q$, except for the initial points of 16 and 11 nm samples which show deviations from the fit, mainly due to the low resolution of small size particles in the SAXS experiment.

The correlation length, which indicates the decay length of the ordering, is reminiscent of the Debye screening length. The previous work showed that the decay length of the interaction between two flat surfaces was found as the Debye length

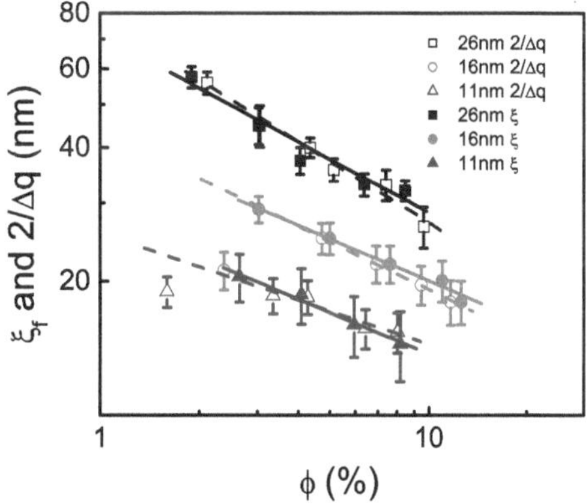

Fig. 4.18 Comparison between AFM decay length ξ_f and SAXS correlation length $2/\Delta q$

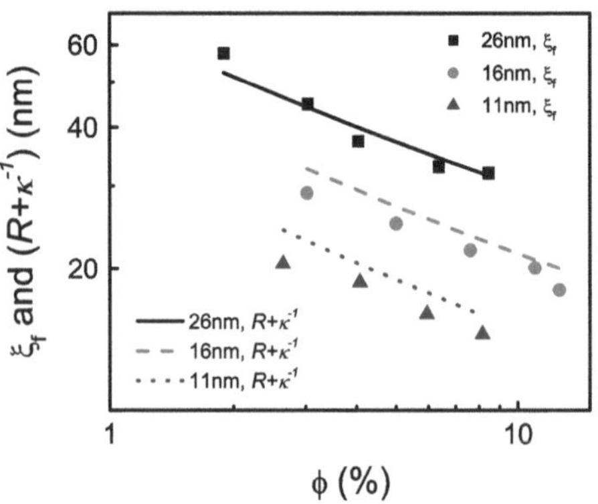

Fig. 4.19 Comparison between AFM decay length ξ_f and the proposed correlation length, $R + \kappa^{-1}$. *Solid line*: $R + \kappa^{-1}$ for 26 nm, *dash line*: $R + \kappa^{-1}$ for 16 nm, *dot line*: $R + \kappa^{-1}$ for 11 nm

[35] The scaling laws in Fig. 4.18 is -0.39, -0.33, and -0.29 for 26, 16 and 11 nm sized particles, respectively. The variance in the scaling law suggests that the particle size has a significant influence on the correlation length. Thus, the expression $\xi = R + \kappa^{-1}$ is herein used for the predicted correlation length between particles. Figure 4.19 shows the comparison of experimental correlation lengths obtained from AFM force curves and the predicted values by assuming $\xi = R + \kappa^{-1}$. These two values coincided with each other for all sized particles.

The radius-subtracted correlation length versus the total ionic strength of the samples I_{tot} is then plotted, where $(2N_A I_{tot})^{-1/2} = (Z\rho + 2N_A I_{salt})^{-1/2} \propto \kappa^{-1}$. The master curve in Fig. 4.20 shows that for all series of particles, the scaling law of radius-subtracted correlation length with ionic strength is remarkably close to $-1/2$. This $-1/2$ scaling law with respect to the ionic strength can be suggested to apply to various systems by excluding the geometries of investigated samples. For specific systems, the ionic strength of the solution is attributed solely by the investigated samples, (e.g. charged colloids and polyelectrolytes in the absence of added salts)

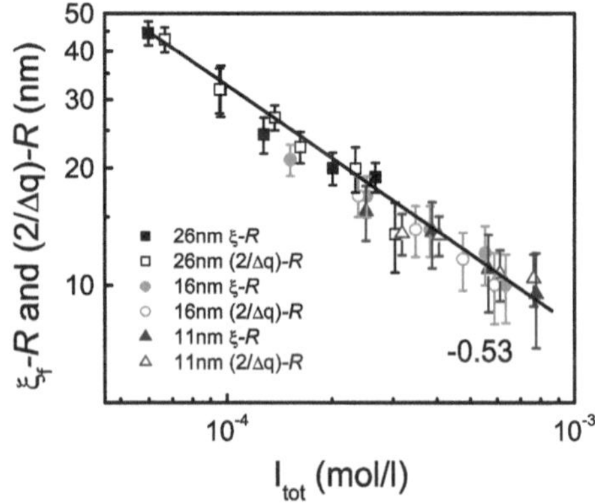

Fig. 4.20 The radius-subtracted correlation length versus the total ionic strength of the sample. The scaling law for all three series of samples is close to $-1/2$, indicating the correlation length is the sum of particle radius and Debye length

I_{tot} is then proportional to the sample concentration, thus $-1/2$ scaling law of ξ-R with sample concentration c can be applied [27].

The consistence of measured ξ with $R + \kappa^{-1}$ indicates the correlation length of the present system is both particle size and ionic strength controlled, in contrast to the negligible influence of the particle size and ionic strength on the interparticle distance. The decrease of the correlation length with increase of particle concentration can be understood as a simple screening effect due to increased ionic strength associated with particle concentration. The particle concentration affects the correlation length through the ionic strength of the total suspensions instead of through the volume scale for interparticle distance.

To illustrate the dependency of theoretical correlation length on the particle size and ionic strength, Fig. 4.8 is converted into the plot of radius-subtracted decay length versus total ionic strength shown in Fig. 4.21, where a scaling law of $-1/2$ is found for HNC and GCMC, consisting with that of experimental ones. MC simulation (not shown in Fig. 4.21) for decay length in bulk, however, yields a deviation in the scaling law due to the aforementioned reasons. Compared with experimental correlation lengths, HNC or GCMC generates smaller values. Similar behavior is shown in Fig. 4.8. These deviations between the experiments and the model predictions may be taken as a hint for an inaccuracy of the choice of model parameters. Indeed, small deviations of model parameters may affect a highly sensitive quantity such as a correlation length much more than the rather robust wavelength values. The same observations that can be sured between experiments and modeling are the decreasing tendency of decay length with increasing particle concentration and particle size, and a significant influence of the ionic strength on the decay length. Regardless of the relative smaller values obtained from the model predictions, the similar exponent, $-1/2$, of the scaling law behavior indicates that the correlation length is a highly

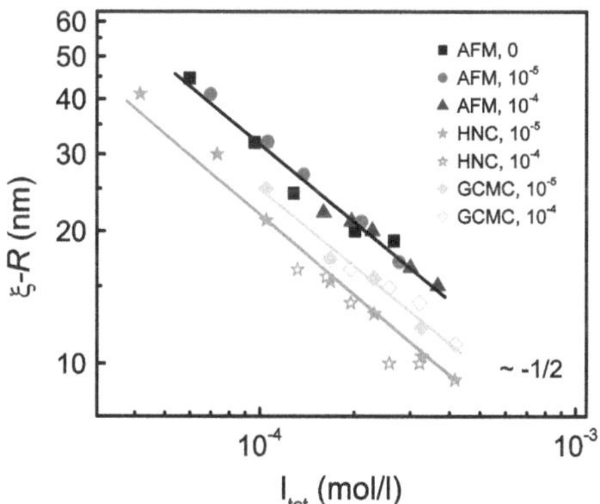

Fig. 4.21 The radius-subtracted correlation length versus the total ionic strength of the sample. The scaling law of HNC calculation for bulk and GCMC simulation for confinement is close to $-1/2$. The difference in the absolute value between experimental and simulated results is due to the uncertainties regarding the choice of asymptotic range

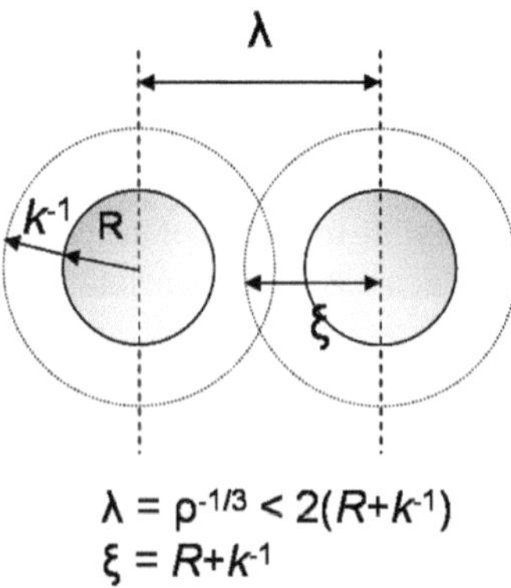

Fig. 4.22 Schematic representation of the relation between λ, ξ and κ^{-1} at concentrations considered in this study

sensitive quantity controlled by particle size and ionic strength of the system. This can be motivated by the fact that on one hand in the low particle concentration regime the range of the correlations is determined by the range of the interaction potential [36]. On the other hand, the range of this potential is determined by the hard-core repulsion with radius R and the DLVO repulsion with range κ^{-1} (see Eq. 2.15). This scaling law is valid up to particle concentrations of 10 vol%. Above this concentration, the systems behave like those with hard repulsive potentials due to the strong screening.

Up to now, the relation of the two characteristic lengths with Debye-Hueckel length can be schematically represented in Fig. 4.22. In the low particle concentration regime, the interparticle distance is always smaller than the effective particle diameter, $\lambda = \rho^{-1/3} < 2(R + \kappa^{-1})$, meaning the diffuse double layers overlap. The correlation length can be proposed as the sum of particle radius and the Debye length, $\xi = R + \kappa^{-1}$.

4.3.4 Dependency of the Particle Interaction Strength

Both SAXS maximum intensity and AFM force amplitude measure the strength of the interactions between particles. As shown in Fig. 4.23a,b, the maximum intensity and force amplitude increases linearly with particle concentration at fixed size. Analogous behavior was also found for confined polyelectrolytes solutions where the amplitude increased with increasing concentration and polymer charge density due to higher overall charge [27]. For larger-sized particles, the increase in the interaction strength was more pronounced. The ratio of the slopes of the maximum scattering intensity versus particle number density in Fig. 4.23a equals to the ratio of the particle size with power law of six, σ^6, indicating that scattering intensity is proportional to the product of particle number density and square of the particle volume due to scattering mechanism ($I_{max} \propto \rho \times V_p^2$). The ratio of the slopes of the force amplitude curves in Fig. 4.23b is equivalent to the ratio of the square of the total charge of the particle

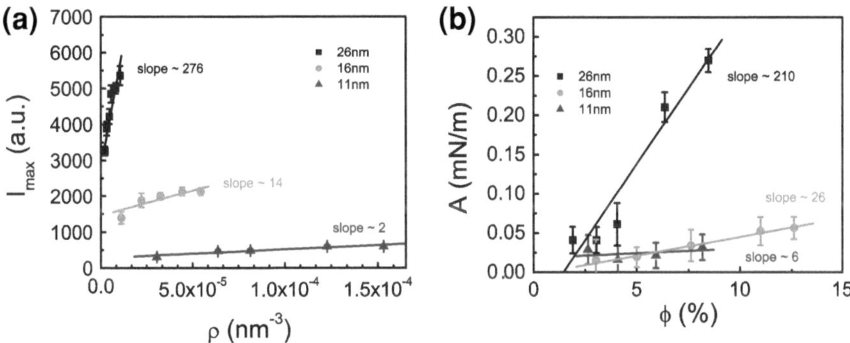

Fig. 4.23 a SAXS maximal intensity versus particle number density. Slope ratio 276: 14: 2 is similar to the square of particle's volume 26^6: 16^6: 11^6. **b** AFM force amplitude versus particle concentration. Slope ratio 210: 26: 6 is similar to the ratio of square of each surface charge 35^2: 13^2: 6^2

Z^2 (Z = 35, 13, and 6 for 26, 16 and 11 nm sized particle, respectively), indicating that the interaction between particles is electrostatic repulsion dominated and particle surface charge influenced (see Eq. 2.15). Previously, one observed that the increase in particle concentration led to an increase in amplitude while the associated increase in counterion concentration led to a decrease in amplitude (Figs. 4.2 and 4.5). The SAXS measurements also showed that, at a given particle concentration, increasing salt concentration caused a reduced intensity (Fig. 4.4). However, the linearly increasing amplitude with no observed maximum in Fig. 4.23b indicates that the effect of particle charge dominates the counterion effect in this study.

There is no direct way to compare the maximum intensity from SAXS with the force amplitude from AFM in order to know how the interaction strength changes in a confined system. The influence of confining surface potential on the structuring of particles is going to be discussed in the next chapter.

4.3.5 Effect of Confinement

So far, the agreement between $\xi - 2/\Delta q$ and $\lambda - 2\pi/q_{max}$ indicates average interparticle distance and correlation length in the direction perpendicular to the confining surfaces correlated well with the bulk ones. No confinement effect in terms of the average interparticle distance and correlation length was observed at the particle concentration considered. The question then is whether there is an effect of confinement on another scale. The occurrence of an oscillatory force itself is a confinement effect, which is caused by the oscillatory density profiles of particles in confinement and represented as layers of particles with varying particle densities formed parallel to the confining surfaces. This confinement induces layering of nanoparticles, in the vicinity of the confining surfaces, indicating that the translational symmetry of the bulk system is broken [2, 3]. At particle concentration below 10 vol%, the particles within the layers are fluid-like as in bulk and the asymptotic range is valid until to the first minimum. This fluid-like in-plane structuring was also addressed by previous experimental and theoretical studies at low particle concentrations [37–39]. Those previous studies also showed that a higher ordering started to form within the contact layer as particle concentration further increased and the full oscillation deviated from the exp()×cos() asymptotic behavior. This is confirmed by the present results on 26 nm sized particles at concentration of 10.9 vol% (shown in Fig. 4.2).

4.4 Conclusion

The dominating wavelengths of the oscillations in characteristic bulk correlation functions and confined charged silica solutions are found to be in excellent agreement with each other, $\lambda_f = \lambda_b$, both from experimental (AFM, SAXS) and theoretical (MC/GCMC, HNC) point of view. The experimental wavelengths are reproduced very well by theoretical calculations based on the DLVO interaction potential. Strictly speaking, the latter is an effective potential derived for bulk systems with spherical counterion distribution. Clearly, this will change in a nanoscopic system where many particles are close to an interface (where image charge effects may also play a role) [40–42]. From that point of view, the good performance of the bulk calculations in the confinement and the agreement between two experimental results indicates that the confinement-induced changes of the wavelength are irrelevant for the quantities considered.

At a fixed particle number density the wavelength of the oscillations turns out to be independent of the particle size, the surface charge of the particles, and ionic strength of the suspensions. Regarding the particle number density dependence of the wavelength, the experimental results reveal an "ideal" scaling behavior described by $\lambda = \rho^{-1/3}$ within the error of the wavelength determined from fits and the error caused by the determination of the particle diameter. This ideal scaling indicates that the wavelength of the confined, layered systems behaves like the average particle distance in an isotropic bulk system. Theory modeling yields very similar results as the experimental ones.

Of course, the oscillatory force is a consequence of the confinement, meaning the translational symmetry is broken. The wavelength λ_f considered in this study is associated with the density distribution perpendicular to the walls, and one has seen that this wavelength strongly and solely depends on the particle number density ρ. Clearly, one would also expect an increase of lateral order with ρ, including the possibility of wall-induced crystallization. Hints for such behavior were already observed in X-ray experiments [43] and also in the present study via a deviation of the measured force $F(h)$ from simple oscillatory behavior in ultrathin films at high particle volume fraction.

In contrast to the wavelength, the dominating correlation length of the oscillations has been found to be not only dependent on the particle number density but also on the particle size and the ionic strength of the suspensions. The increase of the particle concentration, corresponding increase in the ionic strength, and the decrease of particle size lead to the decrease of the correlation length. The relation between the correlation length and Debye screening length can be proposed as $\xi = R + \kappa^{-1}$, meaning the correlation length is both particle size and ionic strength controlled. Theoretical models provide a qualitative agreement with experiments on the correlation length: the dependency of correlation length on the particle size and Debye length has been proven, while the difference in the absolute value between theoretical and experimental results exists in all modeling due to some uncertainties regarding the choice of model parameters. That ξ_f and ξ_b are equal has also been found by AFM and SAXS and is consistent with the prediction from DFT [29].

Both experiment and simulations indicate that increase of particle size/charge leads to both a pronounced increase of the amplitude and the range of the interaction. The AFM force amplitude is proportional to the product of particle volume fraction and square of the particle surface charge, indicating that the particle charge exerts a strong effect on the amplitude because the particle-particle interaction is dominated by electrostatic repulsion. The SAXS maximum intensity is proportional to the product of particle number density and the square of particle volume, due to the scattering mechanism.

References

1. Bechinger, C., Rudhardt, D., Leiderer, P., Roth, R., & Dietrich, S. (1999). *Physical Review Letters, 83*, 3960–3963.
2. Chaudhury, M. (2003). *Nature, 423*, 131–132.
3. Wasan, D., & Nikolov, A. (2003). *Nature, 423*, 156–159.
4. Israelachvili, J., & Pashley, R. (1983). *Nature, 306*, 249–250.
5. Wasan, D., Nikolov, A., Kralchevsky, P., & Ivanov, I. (1992). *Colloids and Surface, 67*, 139–145.
6. von Klitzing, R., & Muller, H. (2002). *Current Opinion in Colloid and Interface Science, 7*, 42–49.
7. Kegler, K., Salomo, M., & Kremer, F. (2007). *Physical Review Letters, 98*, 058304.
8. Stubenrauch, C., & von Klitzing, R. (2003). *Journal of Physics Condensed Matter, 15*, R1197–R1232.
9. von Klitzing, R. (2005). *Advances in Colloid and Interface Science, 114*, 253–266.
10. Ducker, W., Senden, T., & Pashley, R. (1991). *Nature, 353*, 239–241.
11. Nikolov, A., & Wasan, D. (1989). *Journal of Colloid Interface Science, 133*, 1–12.
12. Basheva, E., Danov, K., & Kralchevsky, P. (1997). *Langmuir, 13*, 4342–4348.
13. Sethumadhavan, G., Nikolov, A., & Wasan, D. (2001). *Journal of Colloid Interface Science, 240*, 105–112.
14. Denkov, N., Yoshimura, H., Nagayama, K., & Kouyama, T. (1996). *Physical Review Letters, 76*, 2354–2357.
15. Sharma, A., & Walz, J. (1996). *Journal of the Chemical Society Faraday Transactions, 92*, 4997–5004.
16. Sharma, A., Tan, S., & Walz, J. (1997). *Journal of Colloid Interface Science, 191*, 236–246.

17. Piech, M., & Walz, J. (2002). *Journal of Colloid Interface Science, 253*, 117–129.
18. Piech, M., & Walz, J. (2004). *Journal of Physical Chemistry B, 108*, 9177–9188.
19. McNamee, C., Tsujii, Y., Ohshima, H., & Matsumoto, M. (2004). *Langmuir, 20*, 1953–1962.
20. Tulpar, A., Van Tassel, P., & Walz, J. (2006). *Langmuir, 22*, 2876–2883.
21. Drelich, J., Long, J., Xu, Z., Masliyah, J., Nalaskowski, J., Beauchamp, R., et al. (2006). *Journal of Colloid Interface Science, 301*, 511–522.
22. McNamee, C., Tsujii, Y., & Matsumoto, M. (2004). *Langmuir, 20*, 1791–1798.
23. Milling, A. (1996). *Journal of Physical Chemistry, 100*, 8986–8993.
24. Biggs, S., Burns, J., Yan, Y., Jameson, G., & Jenkins, P. (2000). *Langmuir, 16*, 9242–9248.
25. Biggs, S., Prieve, D., & Dagastine, R. (2005). *Langmuir, 21*, 5421–5428.
26. Qu, D., Baigl, D., Williams, C., Mohwald, H., & Fery, A. (2003). *Macromolecules, 36*, 6878–6883.
27. Qu, D., Pedersen, J. S., Garnier, S., Laschewsky, A., Moehwald, H., & von Klitzing, R. (2006). *Macromolecules, 39*, 7364–7371.
28. Evans, R., Henderson, J., Hoyle, D., Parry, A., & Sabeur, Z. (1993). *Molecular Physics, 80*, 755–775.
29. Grodon, C., Dijkstra, M., Evans, R., & Roth, R. (2005). *Molecular Physics, 103*, 3009–3023.
30. Klapp, S. H. L., Grandner, S., Zeng, Y., & von Klitzing, R. (2008). *Journal of Physics Condensed Matter, 20*, 494232.
31. Kolaric, B., Jaeger, W., & von Klitzing, R. (2000). *Journal of Physical Chemistry B, 104*, 5096–5101.
32. Klapp, S. H. L., Grandner, S., Zeng, Y., & von Klitzing, R. (2010). *Soft Matter, 6*, 2330–2336.
33. Nikolov, A., & Wasan, D. (1992). *Langmuir, 8*, 2985–2994.
34. Klapp, S. H. L., Zeng, Y., Qu, D., & von Klitzing, R. (2008). *Physical Review Letters, 100*, 118303.
35. Sokolov, I., Ong, Q. K., Shodiev, H., Chechik, N., James, D., & Oliver, M. (2006). *Journal of Colloid and Interface Science, 300*, 475–481.
36. Hansen, I. R., & McDonald, J. P. (2006). *Theory of simple liquids* (3rd ed.). Amsterdam: Academic Press.
37. Chu, X., Nikolov, A., & Wasan, D. (1994). *Langmuir, 10*, 4403–4408.
38. Wasan, D., Nikolov, A., & Moudgil, B. (2005). *Powder Technology, 153*, 135–141.
39. Grandner, S., & Klapp, S. H. L. (2008). *Journal of Chemical Physics, 129*, 244703.
40. Netz, R. (2000). *European Physical Journal E, 3*, 131–141.
41. Netz, R. (2001). *European Physical Journal E, 5*, 557–574.
42. Frydel, D., Dietrich, S., & Oettel, M. (2007). *Physical Review Letters, 99*, 118302.
43. Zwanenburg, M., Bongaerts, J., Peters, J., Riese, D., & van der Veen, J. (2000). *Physical Review Letters, 85*, 5154–5157.

Chapter 5
Structuring of Nanoparticles Between Modified Solid Surfaces

5.1 Introduction

Typically, structural force of confined nanoparticles has a damped oscillatory character as a function of the surface separation [1, 2], reflecting the oscillatory density profile, signifying the formation of layers of nanoparticles parallel to the surfaces. In the previous chapter, the *asymptotic* behavior of the structural forces has been demonstrated, particularly the wavelength and decay length of the oscillations at large surface separations are governed by the pair structure in the corresponding bulk fluid. This observation is fully consistent with predictions from density functional theory (DFT) [3, 4], according to which the properties of the surfaces should become irrelevant in the asymptotic limit. On the other hand, DFT also predicts that the surface properties (or, more specifically, the interaction between a charged particle and a surface) do influence the *amplitude* and *phase* of the oscillations.

The properties of the surface are studied in two aspects in this chapter: the surface potential and the surface roughness. The surface potential is modified by depositing a negatively-charged mica sheet on top of a silicon wafer. To understand the mechanisms of the change in structuring after modifying the surface potential, a grand-canonical Monte Carlo simulation (GCMC) involving confined silica particles, which interact via the DLVO potential, is included [5]. The GCMC results only involving silica ions turn out to be highly sensitive with respect to the actual model for the interaction between a silica particle and the surface(s). In particular, the simulated observations are *not* reproduced even qualitatively and predicts an opposite behavior as the experimentally observed one when the simplest version of linearized Poisson-Boltzmann (PB) theory [6, 7] is employed, where the exponential decay of the potential is determined by the bulk Debye screening length and the wall potential only comes into play through a prefactor.

To solve this contradiction, a modified fluid-wall (particle-confining surface) potential is introduced, starting from a PB-like theory for a colloidal suspension next to one charged surface [8]. The modification consists of supplementing the bulk Debye screening length appearing in the simplest approach by a contribution from

Y. Zeng, *Colloidal Dispersions Under Slit-Pore Confinement*, Springer Theses,
DOI: 10.1007/978-3-642-34991-1_5, © Springer-Verlag Berlin Heidelberg 2012

the wall counterions [5]. A similar idea though in a different context is followed in various earlier investigations [9–11]. In these studies, however, the contribution of the wall counterions to the screening parameter in the resulting potential was assumed to be homogeneous. In the present work, at least approximately, the inhomogeneity of the counterion distribution is taken into account, which yields a particle-wall screening length which depends both on the wall counterions (or equivalently, the wall charge) and on the distance between particle and wall. The full fluid-wall potential from the two charged surfaces is then constructed by linear superposition (LSA). The resulting potential is still purely repulsive, but displays a non-monotonic behavior as a function of the wall potential with respect to the degree and range of repulsion. In particular, within the experimentally relevant range of surface potentials, the GCMC results with the new fluid-wall potential model is in qualitative agreement with the experiments.

While oscillatory forces between smooth surfaces are relatively well understood, it is of central importance to understand the effect of surface roughness. Surface roughness is encountered naturally in almost all applicable systems. According to the previous molecular simulation [12, 13] and grand canonical Monte Carlo (GCMC) simulation studies on simple fluids [14–16], the roughness of the confining surfaces is just as important as the nature of the fluid molecules in determining the oscillatory forces. The registry of the confining surface is seen to play an important role on the equilibrium structures formed under confinement and the effect of random roughness reveals a reduction in the oscillatory force amplitude with increasing roughness. At roughness around 33 % of the fluid molecular diameter, the oscillation disappeared.

There are only a few prior experimental studies which have reported the oscillatory forces between surfaces that are no longer atomically smooth. For rough mica surfaces prepared by depositing a compressed Langmuir-Blodgett film of dioctadecyldimethyammonium bromide (DOAB) or adsorbing hexadecyltrimethylammonium bromide (CTAB), the oscillatory force profile of the non-polar liquid octamethylcyclotetrasiloxane measured by surface force apparatus (SFA) showed a reduced range of oscillation (DOAB) or even disappeared altogether (CTAB) [17]. This occurs even if the liquid molecules themselves are perfectly capable of ordering into layers. However, in this scarce experimental study, the influence of surface roughness was not separated from the effect of surface potential. It is not straightforward to study surface roughness since surface potential plays an important role in amplitude and phase of the oscillations [5] and often varies with the roughness.

To address this remaining problem, we apply the layer-by-layer technique in the present study to coat polyelectrolytes on the confining surface, thus modifying the surface properties. The layer-by-layer technique, or consecutive adsorption of oppositely charged polyelectrolytes, has been introduced by Decher et al. [18] This coating is possible because, for many polyelectrolytes, physisorption onto a charged surface is irreversible under mild conditions [19, 20]. Thus, after the first adsorption step the surface can again serve as a substrate for the adsorption of an oppositely charged polyelectrolyte and so on until the desired number of layers is adsorbed. The main features are that the properties of adsorbed polyelectrolyte layers can be easily controlled by the numbers of layers, the ionic strength [21–23], and by the type of polyelectrolyte.

The layer-by-layer technique is applied to a flat silicon wafer with a naturally deposited silica layer on the top, which is used as one confining surface, and to a silica microsphere, when necessary, which is used as the second confining surface. By means of the layer-by-layer physisorption technique, the roughness of the confining surfaces is modified without changing surface potential of a multilayer with the same outermost layer. With increasing number of layers or ionic strength of polyelectrolyte solutions or by selecting appropriate pairs of polyelectrolytes which display similar surface potential, the roughness of the confining surfaces is tuned.

The CP-AFM is then used to measure the oscillatory forces of silica nanoparticle suspensions confined between polyelectrolyte multilayer coated surface(s). The correlation between the amplitude and the phase of the oscillatory force profile of colloidal silica nanoparticles and the roughness of the confining surfaces is investigated. In addition, the effect of surface roughness on the characteristic lengths of the structuring, i.e., the inter-particle distance and the particle correlation length, is studied in this chapter.

5.2 Results and Discussion

5.2.1 Potential of the Confining Surface

In the force experiments two types of substrates are considered: (i) a silicon wafer with a native silica (SiO_2) top layer, and (ii) a freshly cleaved mica sheet deposited on top of a silicon wafer.[1] The corresponding surface potentials are $\psi_S \approx -80\,\mathrm{mV}$ for silica and $\psi_S \approx -160$ mV for mica, respectively. The CP-AFM results for the force-distance curves of 26 nm sized silica particle suspensions, $F(h)$, involving two different (silica and mica) surfaces are presented in Fig. 5.1. One immediately sees that the larger (absolute) surface potential related to the mica surface leads to a pronounced enhancement of the oscillations as compared to the silica surface. To quantify the effect force curves have been fitted according to Eq. 2.14. Results for the amplitude A, wavelength λ_f, and correlation length ξ_f are given in Table 5.1.

The data show that the amplitude A obtained for the (more strongly charged) mica surface is nearly twice as large as that for silica. On the other hand, the wavelength λ_f and correlation length ξ_f of the oscillations remain essentially unaffected. Similar results were previously observed for confined polyelectrolytes [24]. From a conceptual perspective, the constant behavior of λ and ξ suggests that the characteristic lengths are determined rather by the pair structure *among* the particles than by their interaction with the wall. Indeed, the experimental observation that the

---

[1] Reprinted with permission from: *Impact of surface charges on the solvation forces in confined colloidal solutions*, Stefan Grandner, Yan Zeng, Regine von Klitzing, and Sabine H. L. Klapp, *The Journal of Chemical Physics*, **2009**, *131*, 154702. Copyright (2009), American Institute of Physics.

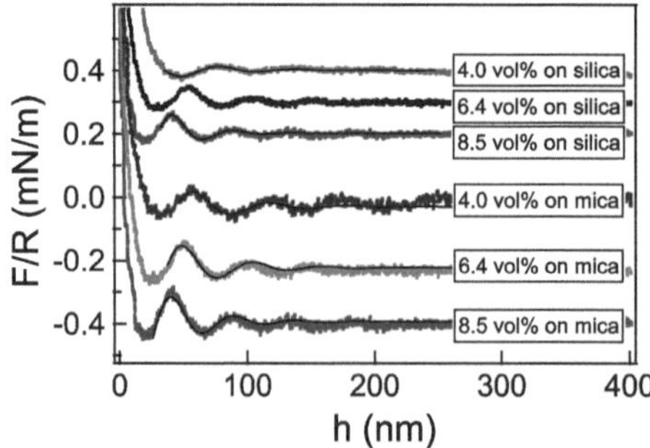

Fig. 5.1 AFM force curves of 26 nm sized silica particle suspensions confined between a silica micro-sphere (on an AFM cantilever) and a silica (*top*) and mica (*bottom*) surface characterized by surface potentials $\psi_S \approx -80$ and -160 mV, respectively. The data have been vertically shifted for ease of viewing. The *solid lines* are fits according to Eq. 2.14. Three concentrations are represented on each surface

Table 5.1 Surface potential ψ_S (-80 mV corresponds to silica and -160 mV corresponds to mica), Amplitude A, wavelength λ_f, and decay length ξ_f of $F(h)$ for different particle volume fraction as obtained from the CP-AFM in Fig. 5.1

ϕ (vol%)	ψ_S (mV)	A (mN/m)	λ_f (nm)	ξ_f (nm)
4.0	-80	0.06	59.2	37.4
4.0	-160	0.11	60.3	36.8
6.4	-80	0.21	50.3	33.0
6.4	-160	0.45	51.0	33.1
8.5	-80	0.27	45.5	32.1
8.5	-160	0.59	45.5	31.5

surface potential influences the amplitude but not the characteristic lengths are fully consistent with rigorous predictions from DFT [4].

Clearly, such an enhancement in oscillation amplitude can arise due to various mechanisms, including the possibility that more particles move from the connected bulk reservoir into the slit. Indeed, such a situation has recently been observed in an investigation of charged colloids in a charged wedge [11], where the colloids turn out to accumulate in the cusp due to a localized, attractive region in the interaction potential between a colloid and the walls. Another possible explanation for the present observation is that the increase of wall potential strongly enhances the Coulomb repulsion between the silica particles and the like-charged wall, leading to a stronger layering of particles inside of the slit.

To understand the underlying mechanisms, a grand-canonical Monte Carlo simulation of a coarse-grained model involving confined silica particles, which interact via the DLVO potential, is included as well [5]. Firstly, the GCMC simulation results based on the simplest model for the fluid-wall interaction which neglects the effect of wall counterions on the screening are briefly considered (Eq. 2.23). Corresponding numerical data for the normalized normal pressure $f(L_z) = P_{zz} - P_b$ as a function of the wall separation L_z (same as h used in the AFM force curves) and the (negative) surface potential ψ_S are presented in Fig. 5.2.

All functions $f(L_z)$ display the damped oscillatory behavior, with the oscillations vanishing upon reaching the bulk limit $L_z \rightarrow \infty$ (i.e., $P_{zz} \rightarrow P_b$). More significant in the present context, however, is the fact that the amplitude of $f(L_z)$ (and thus, the amplitude of the force) decreases monotonically upon increase of $|\psi_S|$. This clearly *contradicts* the AFM experimental results. From a theoretical point of view, the behavior of $f(L_z)$ is a direct consequence of the corresponding behavior of the fluid-wall potential which becomes progressively more repulsive upon increase of $|\psi_S|$. Thereby more and more particles are expelled from the slit. This is also reflected by the GCMC results for the mean silica density, $\bar{\rho}$, plotted in the inset of Fig. 5.2: at fixed L_z, $\bar{\rho}$ becomes smaller the more negative ψ_S is. Indeed, the slit becomes essentially empty at small L_z already at $\psi_S = -27.7$ mV, a wall potential far below that characteristic of a real silica surface.

Having in mind these (obviously wrong) predictions, the corresponding GCMC results based on a new fluid-wall potential, $u_{FS}(z)$ (Eq. 2.26) is considered in Fig. 5.3, where the screening parameter depends on ψ_S and is space-dependent. This is motivated by the release of additional (wall) counterions which accumulate at the walls (Eq. 2.25). Clearly, the dependence of the functions $f(L_z)$ and $\bar{\rho}(L_z)$ on ψ_S is non-monotonic. The extracted parameters, P_{max} (height of the first maximum), θ_f (phase), λ_f (wavelength), and ξ_f (decay length) as a functions of ψ_S were listed in the corresponding paper [5] upon fitting the curves with Eq. 2.21.

When "switching on" the surface potential from $\psi_S = 0$ up to a value of about $|\psi_S| = 40$ mV the quantity P_{max} first decreases. Upon further increasing $|\psi_S|$ towards 80 mV and 160 mV (which are the experimentally relevant values for silica and mica, respectively), P_{max} increase. It is interesting in this context that the value of $|\psi_S| = 40$ mV where P_{max} changes its behavior corresponds to the "reversal point" of the fluid-wall potential. Similar to P_{max}, an increase of the maximum of the corresponding force-distance curves $F(L_z)$ obtained by integration of $f(L_z)$ can be observed (see Fig. 5.3c). Thus, GCMC simulations with a modified fluid-wall potential reproduce,

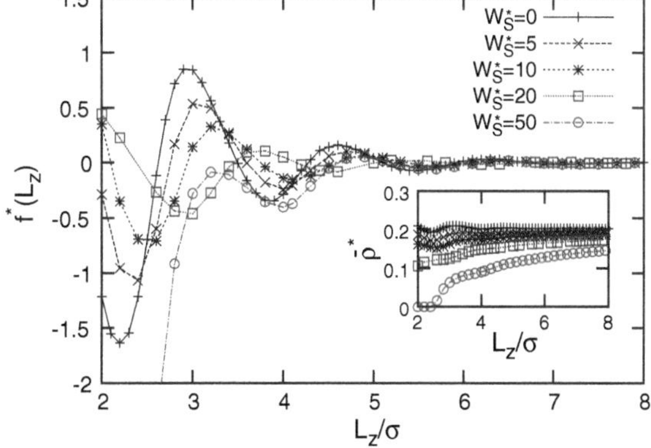

Fig. 5.2 Dimensionless solvation pressure $f^*(L_z) = P_{zz}^* - P_b^*$ for ϕ=10.5 vol% and various surface potentials ψ_S as calculated by GCMC simulations involving the simplest model. The values of $W_S^* = \beta W_S$ correspond to $\psi_S = 0, -2.7, -5.4, -10.9,$ and -27.7 mV, respectively. The inset shows the corresponding mean silica particle density $\bar{\rho}^*$ as a function of the wall separation

Fig. 5.3 GCMC results for **a** the reduced solvation pressure and **b** the mean pore density $\bar{\rho}$ at surface potentials $\psi_S =$ 0, -40, -80 (*silica*), 120, and -160 mV (mica). The corresponding bulk concentration is $\phi=10.5$ vol%. The *solid lines* are fit functions obtained from Eq. 2.21. **c** Shows the resulting structural forces $F(L_z)/2\pi R$ for $\psi_S = -80$ (*dashed*), -120 (*dotted*), and -160 mV (*dot-dashed*). For clarity the curves in (**a**) are shifted along the y-axis

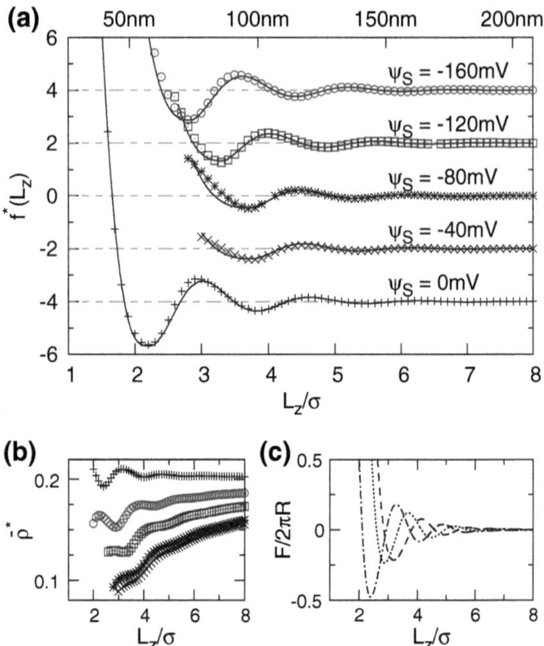

on a qualitative level, the charge-induced enhancement of the oscillations observed in the CP-AFM experiments.

The potential effects on the solvation pressure are mirrored by corresponding effects on the mean density $\bar{\rho}(L_z)$ of the silica particles, which is plotted in Fig. 5.3b. In particular, within the experimentally relevant range of 80 mV $\leq |\psi_S| \leq 160$ mV, the density at a fixed separation L_z increases with $|\psi_S|$, which is consistent with the enhancement of pressure oscillations. This enhancement of the mean particle density reveal that more particles move into the slit with increasing the surface potential. This is due to the corresponding decrease in the range of particle-wall interaction, resulting from the additional contribution of the wall-counterions into the Debye length. At a given wall separation, the shorter the particle-wall interaction range the more layers of particles can fit into the slit.

The changes in $P_{\max}$ and $F_{\max}$ with ψ_S are accompanied by the changes in the phase shift, θ_f. The latter displays a maximum at $\psi_S \approx -40$ mV. In general, the phase shift can be also considered as the depletion zone, which is the separation between the contact layer of particles and the wall. The decrease of θ_f for $|\psi_S| \geq 40$ mV is interpreted as a consequence of the corresponding decrease in the range of $u_{FS}(z)$. The less the particle-wall interaction range the narrower the depletion zone.

On the other hand, the wavelength λ_f remains essentially constant when ψ_S is changed, in agreement to the experimental observations. The correlation length ξ_f varies only slightly, given the difficulties to obtain accurate values for this quantity (i.e., the large error bars). Nevertheless ξ_f is judged to remain essentially unaffected as

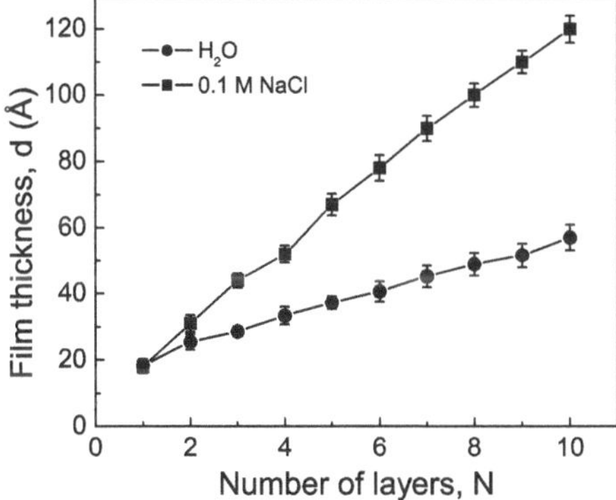

Fig. 5.4 Ellipsometry measurements on film thickness as a function of the number of layers of polyelectrolyte adsorbed on the silicon wafer, i.e., si-PEI-PSS-(PAH/PSS)$_n$, either with 0 or 0.1 M NaCl. n is the number of (polycation/polyanion) double layers and N is the number of total layers. $N = 1$ stands for only one layer of PEI, even number of N stands for PSS as outermost layer, while odd number of N, expect $N = 1$, stands for PAH as outermost layer

well. This is indeed what one would expect based on theoretical arguments: according to DFT, the precise nature of fluid-wall interactions does influence the amplitude and phase of the (asymptotic) pressure oscillations, but not their wavelength and decay length [4].

5.2.2 Roughness of the Confining Surface

5.2.2.1 Multilayer Characterization

To investigate the effect of roughness of confining surfaces on the ordering of nanoparticles, the silicon wafers and silica microspheres were modified by physisorption of two oppositely charged polyelectrolytes one by one, e.g., PAH and PSS. A layer of PEI was pre-adsorbed onto the silica surface for stabilizing the later adsorption [25].[2] Polyelectrolyte concentration was kept at 10^{-2} monoM.

In order to ascertain that each polyelectrolyte had been adsorbed successfully onto the surface, ellipsometry measurements were made to characterize the film thickness grown on the silicon wafers. Figure 5.4 shows a regular growth of the thickness of PAH/PSS multilayer obtained from ellipsometry measurements at assembling salt concentration of 0 and 0.1 M NaCl, respectively, confirming the success of consecutive assembling. N refers to the number of total layers. In the salt-free case, the

---

[2] Reprinted with permission from: *Oscillatory Forces of Nanoparticle Suspensions Confined between Rough Surfaces Modified with Polyelectrolytes via the Layer-by-Layer Technique*, Yan Zeng, Regine von Klitzing, *Langmuir*, **2012**, *28*, 6313–6321. Copyright (2012) American Chemical Society.

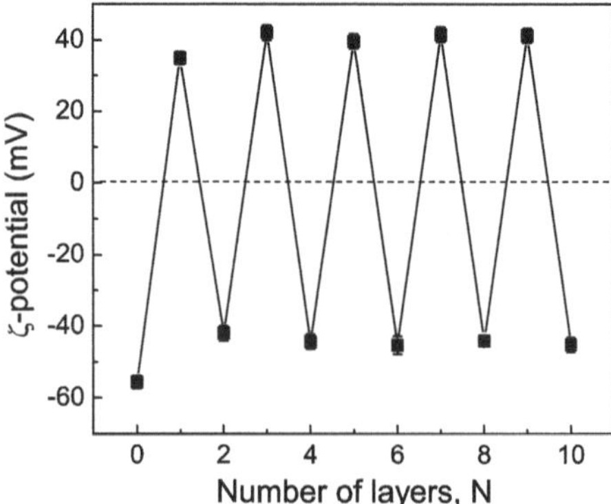

Fig. 5.5 Zeta-potentials of AFM silica microspheres (R = 3.35 µm) assembled by PAH/PSS multilayer prepared in 0 M NaCl. Measurements were carried out in Milli-Q water. The pH value of the solution was around 6.5

growth of the thickness was linear with an average thickness of 4.6 Å per pair of layers, while at the assembling salt concentration of 0.1 M NaCl, the growth of multilayer film was almost linear with an average thickness of 11 Å per pair of layers. It is obvious that after adding extra NaCl into the polyion solution during multilayer assembling, the thickness increases significantly. Adding salt to polyelectrolyte solution during the multilayer assembling can result in a strong screening of the segment charge on the polyelectrolyte chain and thus cause a coil conformation of the complexes [21–23, 26]. These results are consistent with the layer growth reported for PAH/PSS systems [27].

Zeta-potential measurements were performed on the silica probe assembled by 10^{-2} monoM of PAH/PSS with PEI as the first layer. The zeta-potential changed from −55 to +35 mV after PEI was adsorbed and then oscillated between −44 mV for PSS and +41 mV for PAH (Fig. 5.5). The charge reversal confirmed the success of consecutive assembling in each step. The unchanged zeta-potential of PSS- or PAH-ended multilayers, irrespective of the number of assembled layers, indicates that the surface potential of polyelectrolyte adsorbed surfaces does not change with the number of layers [28–32].

The surface potentials of polyelectrolyte-assembled silicon wafers were determined by measuring the approaching part of the corresponding force curve (in "pure repulsion type" and "repulsion and adhesion type", see later section "Force profiles in the absence of nanoparticles") in Milli-Q water. The force curves were fitted with DLVO analysis, in which the electrostatic repulsions were calculated by solving numerically the non-linear Poisson-Boltzmann equation for two parallel surfaces, by assuming constant surface charge or constant surface potential [33].

Generally, fitting with constant surface charge or surface potential is in good agreement with experimental force at larger separations. At small distances the data is better approximated by the constant surface charge model (Fig. 5.6). The surface potential, or the surface charge, hardly changes with increasing number of PAH/PSS multilayer, as long as the multilayer is terminated with the same polyelectrolyte. The average value of surface potentials of PSS terminated multilayers was taken from

Fig. 5.6 The approaching part of a normalized force curve between two bare silica surfaces (AFM silica microsphere and silicon wafer) in Milli-Q water. The best fit at constant surface potential (CP: *dotted line*) and constant surface charge (CC: *solid line*) are shown

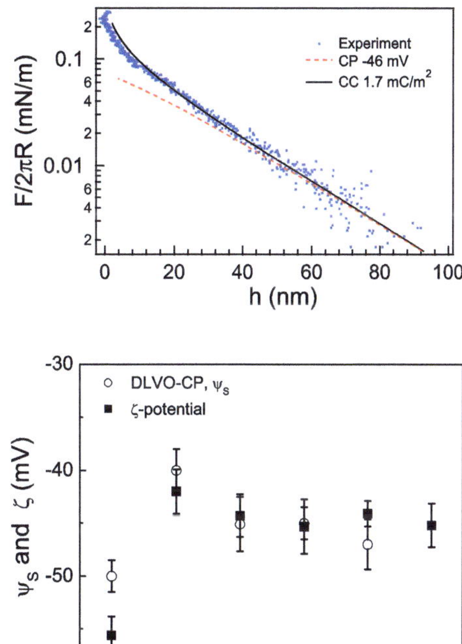

Fig. 5.7 The comparison of the surface potentials extracted from two methods for PAH/PSS multilayers prepared in 0 M NaCl: zeta-potential measurements (*squares*) and DLVO force analysis with constant potential (*circles*)

multiple measurements on the same spot and also on different spots and remained around $-45\,\mathrm{mV}$, while the average value of surface charges was around $2.0\,\mathrm{mC\,m^{-2}}$. No significant difference has been found between the values obtained from curves of "pure repulsion type" and "repulsion and adhesion type". The calculated surface potential agrees with the zeta-potential study on silica microspheres within the margin of error (Fig. 5.7), indicating that the two methods measure the potential at a similar distance from the surface. A difference only exists for bare silica surfaces, which is probably due to the different cleaning process used for spherical and planar surfaces. The same DLVO analysis was also applied on silicon wafers coated with PAH/PSS multilayer in the presence of 0.1 M NaCl during assembly and on one layer of PAA and HA assembled silicon wafers, i.e. si-PEI-PAA and si-PEI-HA (see Table 5.2). The surface potential of PAH/PSS with 0.1 M NaCl shows no significant difference in comparison to that of the salt-free case. The average value is $-44\,\mathrm{mV}$ and stays constant with increasing number of layers. The calculated surface potentials of PAA and HA are -40 and $-42\,\mathrm{mV}$, respectively, which are similar to the previously reported values for PAA and HA from zeta potential measurements [34–36] and also to the aforementioned surface potential of PSS.

The surface roughness is further analyzed from tapping-mode AFM images in aqueous medium at each step with PSS forming the outermost layer. AFM topography images of PAH/PSS multilayer coated silicon wafers immersed in Milli-Q water with scan size 2.5 μm × 2.5 μm are shown in Fig. 5.8. With increasing number of adsorbed layers, the contrast on the surfaces becomes more significant, indicating that the surface roughness increases. The roughness was calculated as a root mean square value over each 1.0 μm ×1.0 μm box in images. The dependency of surface roughness of PAH/PSS multilayer with number of layers is shown in Fig. 5.9. The roughness of PEI-PSS coated silicon wafer is around 12 Å and increases to 21 Å for 10 layers of polyelectrolyte coated silicon wafers. A similar trend of increasing roughness with the number of multilayer up to 10 bilayers has been reported previously [37–39].

The roughness increases with increasing ionic strength of the polyelectrolyte solutions, especially after four layers. The roughness of silicon wafer coated with six layers of polyelectrolyte prepared in 0.1 M NaCl is 19 Å, being close to the highest roughness obtained in the salt-free case. An increase in roughness with salt concentration coincides with previous reports [40, 41]. The roughness of PEI-PAA and PEI-HA coated silicon wafers are 35 and 60 Å, respectively. The higher roughness

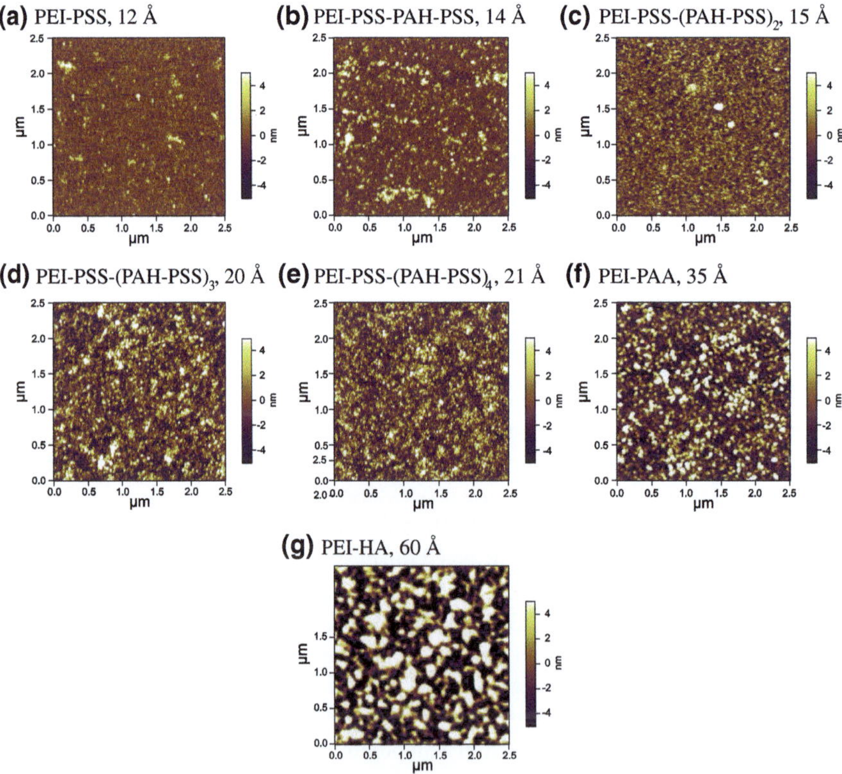

Fig. 5.8 AFM height images of **a–e** PAH/PSS multilayer, **f** PEI-PAA, and **g** PEI-HA adsorbed on silicon wafers obtained with tapping mode in Milli-Q water. The scan sizes are 2.5 × 2.5 μm with a fixed vertical scale of 10 nm. The number in each caption refers to the root mean square roughness

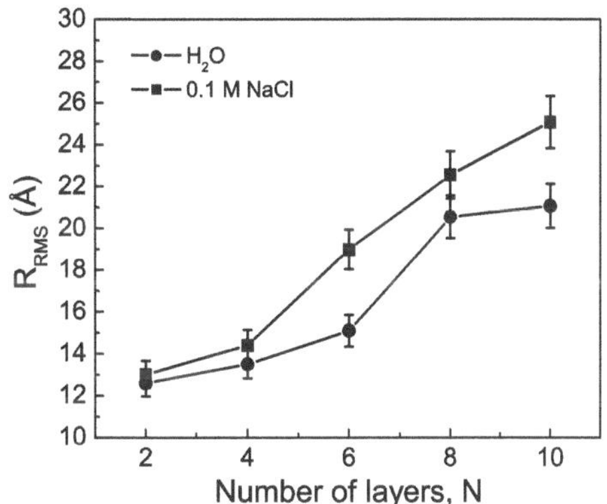

Fig. 5.9 Root mean square roughness R_{RMS} of PAH/PSS multilayer films adsorbed on silicon wafers as a function of the number of layers prepared in 0 and 0.1 M NaCl. The roughness was calculated from $1.0\,\mu m \times 1.0\,\mu m$ boxes on the AFM height images (in Milli-Q water, tapping mode)

Table 5.2 Summary of the surface potentials ψ_S determined by DLVO-analysis with constant potential model for various polyelectrolyte coated silicon wafers

Surface	ψ_S (mV)	
	0 M NaCl	0.1 M NaCl
PEI-PSS	−50	−46
PEI-PSS-PAH-PSS	−40	−44
PEI-(PSS-PAH)$_2$-PSS	−45	−45
PEI-(PSS-PAH)$_3$-PSS	−45	−44
PEI-(PSS-PAH)$_4$-PSS	−47	−48
PEI-PAA	−40	–
PEI-HA	−42	–

is related to a coil conformation of the polyelectrolytes, or more precisely the charge density of the polyelectrolytes (see Discussion).

5.2.2.2 Force Profiles in the Absence of Nanoparticles

The force profiles were taken as a series of measurements with the same silica microsphere on different spots of the same silicon wafer in Milli-Q water. Normally, 10 curves were acquired at each spot and 100 curves in total on a single surface. Three main types of force curves were observed between confining surfaces, based on the number of polyelectrolyte multilayer adsorbed on the surfaces. Figure 5.10a demonstrates the "pure repulsion type", in which both approach and retraction branches show only repulsion between the silica microsphere and silicon wafer surface. In Fig. 5.10b, a weak adhesion appears in the retraction branch, while the approach part remains repulsive. This is referred to as the "repulsion and adhesion type". In Fig. 5.10c, an attraction or no repulsion in the approach branch can also be observed in addition to an adhesion in the retraction branch. This is referred to as the "pure adhesion type".

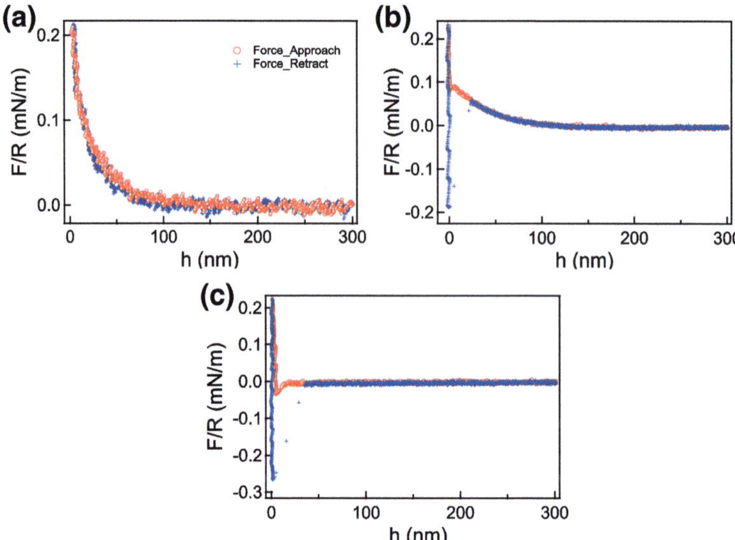

Fig. 5.10 Three types of force curves between a silica microsphere and a silicon wafer coated with polyelectrolyte multilayer in Milli-Q water. **a** Pure repulsion type: complete coverage of polyelectrolyte on the surfaces, no detachment of polyelectrolyte. **b** Repulsion and adhesion type: complete coverage, polyelectrolyte chains get entangled. **c** Pure adhesion type: attraction between two oppositely charged patches due to the detachment of polyelectrolyte or incomplete coverage of polyelectrolyte on the surfaces

The reversible transition between "pure repulsion type" to "repulsion and adhesion type" curves can occur in the same experiment. This transition was observed both on the same spot and on the different spots of the silicon wafer. These two types of force curves are most likely observed on surfaces coated with a larger number of polyelectrolyte layers. In the case of only a few adsorbed layers, the "pure adhesion type" curve can be observed occasionally and the transition back to the other two types is irreversible. It is only possible for the samples which show attraction or no repulsion to restore repulsive behavior upon re-dipping the silicon wafers and/or silica microsphere into the last adsorbed polyion solution.

The same repulsive behavior in the approach branch of the "pure repulsion type" and "repulsion and adhesion type" curves and their coexistence in the same experiment indicate that the corresponding polyelectrolytes do not detach from the surface. If some polyelectrolyte chains transferred from one surface to the other, a partial charge reversal would be expected, resulting in a consequent reduction or annihilation of the repulsive force. The surface potential determined using DLVO analysis shows that there is no significant difference between the values obtained from the "pure repulsion type" and "repulsion and adhesion type" curves, which proves that the surface charge remains constant and no detachment occurs. The observed weak adhesion upon retraction therefore does not have an electrostatic origin. Instead, weak adhesion seems to arise when adsorbed polyelectrolyte chains from the opposing surfaces get entangled. The adhesion takes place in the retraction branch, due to

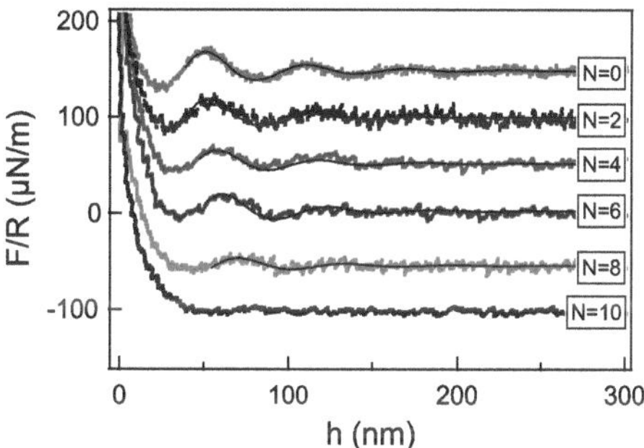

Fig. 5.11 The normalized force curves of 4.0 vol% of 26 nm nanoparticle suspensions confined between a bare AFM silica microsphere and a PAH/PSS multilayer coated silicon wafer. N represents the number of total polyelectrolyte layers with PEI as preadsorbed layer and PSS as the outermost layer

the detachment of the silica microsphere from the silicon wafer and to the bridging and extension of the accompanied polyelectrolyte chains [30].

In Fig. 5.10c, an attraction or no repulsion on the approach branch and no spontaneous transition to a repulsive type curve in a measurement indicate that a partial charge reversal indeed occurs at the surfaces. This type of force curve is mostly due to the detachment of polyelectrolyte chains from one surface to other or the incomplete coverage of the surfaces, which results in a strong interaction of the polycation on one surface with the polyion on the opposing one (briefly, the attraction between two oppositely charged patches). This detachment or incomplete coverage mechanism is confirmed by the fact that this type of force only appears for the first few layers of a coated surface and disappears upon increasing the number of layers. For DLVO surface potential determination, the approaching branch of the "pure repulsion type" and "repulsion and adhesion type" curves should be used and not "pure adhesion type" curves.

5.2.2.3 Force Spectroscopy with Nanoparticles

The force curves of 4.0 vol% of 26 nm silica particle suspensions measured between a bare AFM silica microsphere and a silicon wafer assembled with varying number of PAH/PSS multilayer are shown in Fig. 5.11, where N represents the number of total polyelectrolyte layers with PEI as preadsorbed layer and PSS as the outermost layer. It is obvious that as more polyelectrolytes adsorb onto the silicon wafer, the amplitude of the oscillatory force decreases and the position of first maximum (referred to as the "phase") gradually shifts to larger separations. By fitting the oscillatory force with Eq. 2.14, constant wavelength λ and decay length ξ are obtained, which remain the same as in the case of bare silicon wafer, indicating that the inter-particle distance and the correlation length, respectively, are not affected by the surface roughness. These observations are consistent with the findings of density function theory (DFT) [4].

Figure 5.12 shows the comparison of force curves of 4.0 vol% of 26 nm silica particle suspensions using a bare silica microsphere, both in the presence and absence of salt during preparation. In the case of one layer of PSS coated silicon wafer,

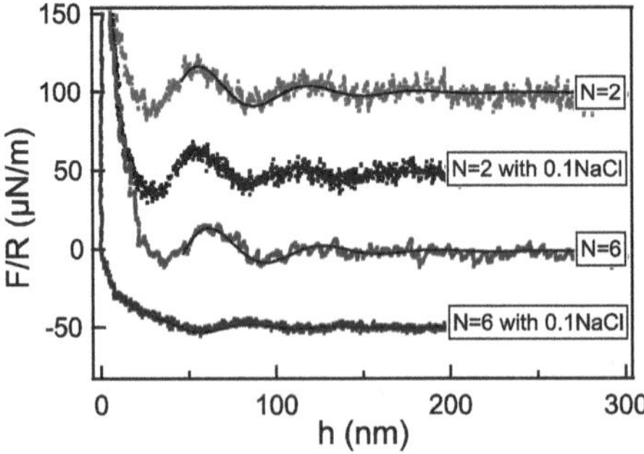

Fig. 5.12 Comparison of force curves of 4.0 vol% of 26 nm silica particle suspensions on PAH/PSS coated silicon wafers in the presence and absence of salt during the polyelectrolyte preparation, using a bare silica microsphere

prepared with 0.1 M NaCl, the oscillation has almost the same amplitude as obtained from that without salt during assembling. In contrast to this, the oscillation in the case of PEI-PSS-(PAH-PSS)$_2$ assembled silicon wafer, with PSS and PAH both prepared in 0.1 M NaCl solution, has a significantly smaller amplitude compared to the oscillation obtained from the silicon wafer coated with same number of layers in absence of salt during the preparation. A phase shift of the experimental curves toward larger separations accompanied the reduction in force oscillation.

Reduced oscillatory amplitude is not only observed upon increasing the number of layers or adding salt during assembly, but it also occurs upon assembling the second confining surface, i.e., the silica microsphere, with PAH/PSS multilayer. The additional influence of coating the second confining surfaces on the force profiles of 4.0 vol% of 26 nm silica particle suspensions is shown in Fig. 5.13, where the oscillatory forces between a bare silica microsphere and PAH/PSS multilayer coated

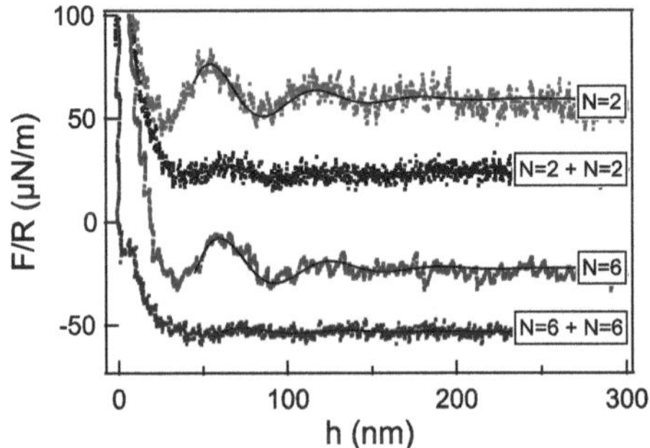

Fig. 5.13 Comparison of force curves of 4.0 vol% of 26 nm silica particle suspensions in respect to bare silica microsphere or PAH/PSS multilayer coated silica microsphere on the silicon wafers coated with N number layers of PAH/PSS multilayer. $N + N$ means that the silicon wafer and silica microsphere both are coated with N number of layers

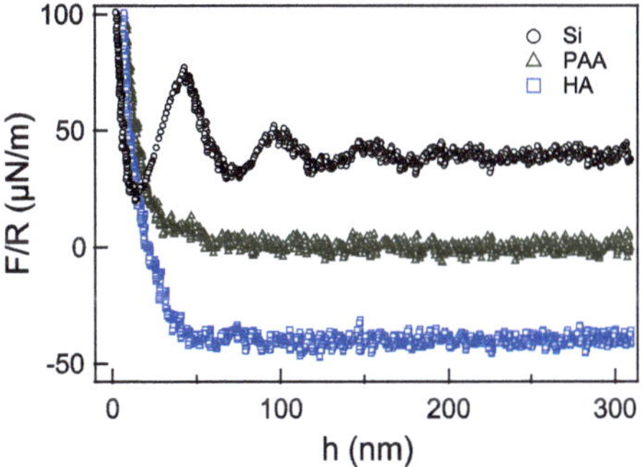

Fig. 5.14 Comparison of surface forces of 4.0 vol% of 26 nm silica particle suspensions between a bare silicon wafer and PEI-PAA and PEI-HA coated silicon wafer, using a bare silica microsphere

silicon wafers are compared with the forces obtained from both surfaces coated with PAH/PSS multilayer. The extra adsorption on the second surface leads to a further reduction in the oscillation amplitude and the oscillation vanishes faster than in the case where just one surface has been modified.

When polyanions PAA and HA were used instead of PSS, the surface forces showed pure monotonic behavior already on silicon wafers coated with only one layer of PAA or HA. Figure 5.14 shows the comparison of surface forces of 4.0 vol% of 26 nm silica particle suspensions on bare silicon wafer and PEI-PAA and PEI-HA coated silicon wafer, using a bare silica microsphere.

5.2.2.4   Force Amplitude Correlation to Surface Roughness

According to previous studies [22, 23, 27], the charge density of the polyelectrolyte plays a very important role in the thickness and roughness of the multilayer. In general, reducing the charge density promotes a coiled polymer chain conformation. The pKa of PAA is about 6.5 [42, 43] which is close to the pH of the assembling solution. Thus the charge density of PAA is assumed to be lower than that of PAH, which has a pKa of 8.8 [43], and of PSS which is a strong polyanion and charged throughout a large pH range. HA has the lowest charge density among these polyelectrolytes. The distance of chargeable groups is 10 Å compared to 2.5 Å in the case of PAH, PSS and PAA. The increased thickness and roughness with increasing ionic strength of the polyelectrolyte solution is also due to the reduced charge density on the polyelectrolyte chains, resulting from the charge screening by the counterions [21, 26]. The layer-by-layer technique can therefore effectively tune the surface roughness with respect to the effective polymer charge.

It's interesting to note that, although the polymer charge density is different for different types of polyelectrolytes at pH 6.5, the measured zeta potential and/or DLVO-analyzed surface potential do not show a significant difference. This might be due to the fact that the shear plane in zeta potential measurements and/or the constant potential range used in calculation are beyond the length scale which is important for observing difference in the surface potentials. Even if there were local variation in surface charges of the outermost layers, they would be compensated by counterions within length scales shorter than the distance of the shear plane and/or constant surface potential range.

In the present work, the reduction in amplitude and shift in phase of the oscillations occurs in four cases: (1) increasing the number of PAH/PSS multilayer, (2) increasing ionic strength of polyelectrolyte solutions during assembly, (3) replacing the bare silica microsphere with one coated with PAH/PSS multilayer, and (4) by adsorbing a layer of PAA or HA onto the silicon wafer instead of PAH/PSS layers. One can exclude the surface potential as dominant factor for the suppression in oscillation in the present study because our results show no change in potential in the above-mentioned cases. Thus the dominant factor of the force damping and phase shift is the increased surface roughness.

The oscillation of the force profile of nanoparticles vanished around the roughness threshold of 21 Å for a single coated confining surface, shown in Fig. 5.11. The effect of roughness of both confining surfaces contributed to the force profiles. This was confirmed by the significant reduction in the oscillation when the second coated surface was introduced (Fig. 5.13). The significant damping in the oscillation at six layers of polyelectrolyte prepared from 0.1 M NaCl in comparison to the negligible change at two layers, shown in Fig. 5.11, correlated with the significantly increased roughness at six layers and negligibly changed roughness at two layers, shown in Fig. 5.9.

At the roughness threshold, the surface topography presented sufficient difference in varied separation distance to smear out the oscillations and gave instead a pure monotonic behavior. In order to show an oscillatory force, the nanoparticles must be able to show positional correlation over a reasonably long range perpendicular to the surface and the correlation function should be the same over a larger lateral area. This requires that both the nanoparticles and the surfaces have a high degree of order or symmetry, otherwise the oscillation would not occur. A roughness of a few nanometers on a single surface (2.1 nm), which corresponds to about 10 % of the nanoparticle diameter (26 nm), was sufficient to eliminate the oscillatory force.

The reduction of the oscillation amplitude and the phase shift toward larger separations correlate with the results obtained by grand canonical Monte Carlo simulation on a Lennard-Jones fluid confined between two rough surfaces [14]. The roughness-induced reduction in amplitude and change in phase can be modeled as a superposition of oscillatory forces between many smooth surfaces lying at slightly different distances. In addition, the simulation result showed the oscillation vanished at surface roughness around 33 % of the fluid molecular diameter. This is larger than the approximate 10 % of the nanoparticle diameter as the roughness threshold to eliminate the oscillatory force of the silica nanoparticles. The silica nanoparticles supposedly yield greater oscillation amplitude and thus larger roughness threshold, although, the smaller roughness threshold might result from the shear flow in the real experiment and the higher polydispersity of the nanoparticles. The shear flow registered in the real experiments, due to the hydrodynamic friction force that accompanied the approach of two surfaces, would disturb the ordering of nanoparticles in the confinement [44]. The varying surface roughness could also affect the local hydrodynamic friction force [45]. The polydispersity of the silica nanoparticles in contrast to the ideal spherical molecules used in the simulation may be another reason for the reduction in the amplitude [46]. In addition, some other parameters set in the simulation may differ from the ones in the real experiment, such as the surface potential.

Finally, it is worth mentioning that the elasticity of confining surfaces can also influence the ordering of the nanoparticles. Measurements between a rigid silica microsphere and a bubble reveals that the surface elasticity reduces the amplitude of oscillations (see Chap. 6). In the present study, however, the thickness of the multilayer in the most cases was less than 6 nm, thus the effect of surface elasticity was suppressed by the hard silica surface. If there was an effect of surface elasticity, it was manifested as surface force curves that did not increase perpendicularly after the "contact point" but rather deviated from the straight line because of deformation of the surface. This deviation was not observed in the present study even though for the thickest film of 12 nm (PAH/PSS, $N = 10$, at 0.1 M NaCl), indicating that the effect of surface elasticity was negligible.

5.3 Conclusion

The confining surfaces were modified by attaching a mica sheet onto a silica substrate or by physically adsorbing polyelectrolytes onto silica surfaces with the layer-by-layer technique. The enhanced surface potential, or surface charge in the first case results in an increase in the oscillatory force amplitude. The underlying mechanism is the fact that more particles move from the connected bulk reservoir into the slit, indicated by the grand-canonical Monte Carlo simulations with a modified particle-wall potential assuming that the charged walls release additional counterions which accumulated in a thin layer at the surfaces. On the other hand, the wavelength and correlation length which characterize the asymptotic behavior of the oscillation have been shown not to change with the confining surface potential, both by experiment and simulation, in agreement with prediction from density functional theory.

In the second case, the wavelength and correlation length of the oscillation have also been shown to be affected neither by the number of multilayers nor by the pair of the polyelectrolytes. A reduced oscillatory amplitude, however, is observed with increasing the number of multilayers and the ionic strength as well as the charge density of the polyelectrolyte chains. The surface potentials of multilayers have been found not to change with the number of multilayers, ionic strength of the polyelectrolytes solution, or pair of polyelectrolytes (although they change after first layer regarding to the bare silica surface), the reduction in the force amplitude thus correlates with the consequently increased surface roughness. A roughness of a few nanometers on a single surface, which corresponded to about 10 % of the nanoparticle diameter, was sufficient to eliminate the oscillatory force of 26 nm diameter silica nanoparticles.

References

1. Israelachvili, J., & Pashley, R. (1983). *Nature, 306*, 249–250.
2. Stubenrauch, C., & von Klitzing, R. (2003). *Journal of Physics: Condensed Matter, 15*, R1197–R1232.
3. Evans, R., Henderson, J., Hoyle, D., Parry, A., & Sabeur, Z. (1993). *Molecular Physics, 80*, 755–775.
4. Grodon, C., Dijkstra, M., Evans, R., & Roth, R. (2005). *Molecular Physics, 103*, 3009–3023.
5. Grandner, S., Zeng, Y., von Klitzing, R., & Klapp, S. H. L. (2009). *Journal of Chemical Physics, 131*, 154702.
6. Hansen, I. R., & McDonald, J. P. (2006). *Theory of simple liquids* (3rd ed.). Amsterdam: Academic Press.
7. Hansen, J., & Lowen, H. (2000). *Annual Review of Physical Chemistry, 51*, 209–242.
8. Bhattacharjee, S., Elimelech, M., & Borkovec, M. (1998). *Croatica Chemica Acta, 71*, 883–903.
9. Denton, A., & Lowen, H. (1998). *Thin Solid Films, 330*, 7–13.
10. Allahyarov, E., D'Amico, I., & Lowen, H. (1999). *Physical Review E, 60*, 3199–3210.
11. Loewen, H., Haertel, A., Barreira-Fontecha, A., Schoepe, H. J., Allahyarov, E., & Palberg, T. (2008). *Journal of Physics: Condensed Matter, 20*, 404221.
12. Curry, J., Zhang, F., Cushman, J., Schoen, M., & Diestler, D. (1994). *Journal of Chemical Physics, 101*, 10824–10832.
13. Gao, J., Luedtke, W., & Landman, U. (2000). *Tribology Letters, 9*, 3–13.
14. Frink, L., & van Swol, F. (1998). *Journal of Chemical Physics, 108*, 5588–5598.
15. Diestler, D., & Schoen, M. (2000). *Physical Review E, 62*, 6615–6627.
16. Porcheron, F., Schoen, M., & Fuchs, A. (2002). *Journal of Chemical Physics, 116*, 5816–5824.
17. Christenson, H. (1986). *Journal of Physical Chemistry, 90*, 4–6.
18. Decher, G. (1997). *Science, 277*, 1232–1237.
19. Decher, G., & Schlenoff, J. (2003). *Multilayer thin films*. Weinheim: Wiley-VCH.
20. von Klitzing, R. (2006). *Physical Chemistry Chemical Physics, 8*, 5012–5033.
21. Kovacevic, D., Van der Burgh, S., de Keizer, A., & Stuart, M. (2002). *Langmuir, 18*, 5607–5612.
22. Schlenoff, J., & Dubas, S. (2001). *Macromolecules, 34*, 592–598.
23. Steitz, R., Leiner, V., Siebrecht, R., & von Klitzing, R. (2000). *Colloids and Surfaces A, 163*, 63–70.
24. Qu, D., Brotons, G., Bosio, V., Fery, A., Salditt, T., Langevin, D., et al. (2007). *Colloids and Surfaces A, 303*, 97–109.
25. Lowack, K., & Helm, C. (1998). *Macromolecules, 31*, 823–833.
26. Kovacevic, D., van der Burgh, S., de Keizer, A., & Stuart, M. (2003). *Journal of Physical Chemistry B, 107*, 7998–8002.
27. Ruths, J., Essler, F., Decher, G., & Riegler, H. (2000). *Langmuir, 16*, 8871–8878.
28. Sukhorukov, G., Donath, E., Davis, S., Lichtenfeld, H., Caruso, F., Popov, V., et al. (1998). *Polymers for Advanced Technologies, 9*, 759–767.
29. Smith, R., McCormick, M., Barrett, C., Reven, L., & Spiess, H. (2004). *Macromolecules, 37*, 4830–4838.
30. Bosio, V., Dubreuil, F., Bogdanovic, G., & Fery, A. (2004). *Colloids and Surfaces A, 243*, 147–155.
31. Dejeu, J., Buisson, L., Guth, M. C., Roidor, C., Membrey, F., Charraut, D., et al. (2006). *Colloids and Surfaces A, 288*, 26–35.
32. Schwarz, B., & Schonhoff, M. (2002). *Langmuir, 18*, 2964–2966.
33. Chan, D., Pashley, R., & White, L. J. (1980). *Journal of Colloid and Interface Science, 77*, 283–285.
34. Burke, S., & Barrett, C. (2004). *Pure and Applied Chemistry, 76*, 1387–1398.
35. Richert, L., Boulmedais, F., Lavalle, P., Mutterer, J., Ferreux, E., Decher, G., et al. (2004). *Biomacromolecules, 5*, 284–294.

36. Burke, S., & Barrett, C. (2005). *Biomacromolecules, 6*, 1419–1428.
37. Lowman, G., & Buratto, S. (2002). *Thin Solid Films, 405*, 135–140.
38. Soltwedel, O., Ivanova, O., Nestler, P., Mueller, M., Koehler, R., & Helm, C. A. (2010). *Macromolecules, 43*, 7288–7293.
39. Ricotti, L., Taccola, S., Bernardeschi, I., Pensabene, V., Dario, P., & Menciassi, A. (2011). *Biomedical Materials, 6*, 031001.
40. Dubas, S., & Schlenoff, J. (2001). *Macromolecules, 34*, 3736–3740.
41. Losche, M., Schmitt, J., Decher, G., Bouwman, W., & Kjaer, K. (1998). *Macromolecules, 31*, 8893–8906.
42. Roma-Luciow, R., Sarraf, L., & Morcellet, M. (2000). *Polymer Bulletin, 45*, 411–418.
43. Choi, J., & Rubner, M. (2005). *Macromolecules, 38*, 116–124.
44. Kleinschmidt, F., Stutbenrauch, C., Delacotte, J., von Klitzing, R., & Langevin, D. (2009). *Journal of Physical Chemistry B, 113*, 3972–3980.
45. Bonaccurso, E., Butt, H., & Craig, V. (2003). *Physical Review Letters, 90*, 144501.
46. Piech, M., & Walz, J. (2002). *Journal of Colloid and Interface Science, 253*, 117–129.

Chapter 6
Structuring of Nanoparticles Confined Between a Silica Microsphere and an Air Bubble

6.1 Introduction

In the previous two chapters the structuring of silica nanoparticles confined between two rigid surfaces has been investigated. The structuring characteristic lengths, wavelength and correlation length, are determined by the particles-quantities rather than by the confining surface charge or surface roughness. The force amplitude, in other words the interaction strength, is influenced by the confining surface charge and surface roughness as well.

The aim of this chapter is to investigate the influence of the confining surface deformability on the structuring of silica nanoparticles. There are just a few reports of the interaction between colloids, such as micelles or latex particles [1–4], confined between deformable surfaces like in a foam lamella. Typically, a thin film pressure balance (TFPB) is used. The existence of oscillatory forces is detected by a sequence of steps in film thickness. The step size between two adjacent repulsive branches is connected to the layering distance or the oscillatory wavelength. The previous unpublished work of our laboratory [5] shows the step size of silica nanoparticles at low particle concentration regime is always twice the particle diameter, irrespective of the particle concentration. This is different from the AFM results obtained between two solid surfaces, where the oscillatory wavelength scales with particle concentration as an exponent of $-1/3$. Thus the measurements on deformable air/liquid interfaces need to be performed by AFM to compare with the TFPB results.

The first AFM force measurement on deformable surfaces has been reported by Ducker et al. [6]. The interaction between a AFM solid sphere and various deformable surfaces were investigated. A silicon wafer was hydrophobilized with a self-assembled monolayer of octadecyltrichlorosilane (OTS) and then a sheet of mica with a hole with radius of $200\,\mu m$ in the center was placed on the top of the

Reproduced by permission of The Royal Society of Chemistry: *Structuring of colloidal suspensions confined between a silica microsphere and an air bubble*, Yan Zeng, Regine von Klitzing, *Soft Matter*, **2011**, *7*, 5329–5338.

silicon wafer. An air bubble can be thus transferred to the center of the hole from a micro-pipette and be stable for many hours. Butt et al. [7–9] simplified the procedure by using a slide of Teflon as substrate. In certain cases, a hole was digged on the Teflon and connected through a tube with a pump, thus air bubbles can be generated with controlled size through defined pressure. Butt et al. also invented a inversed CP-AFM, [10] where the cantilever probe was immersed in the liquid and approach upwards the air/liquid interfaces, thus the interaction between a solid sphere and deformable air/liquid interfaces can be also determined. Dagastine et al. [11, 12] further applied the technique to measure interaction between two air bubbles or oil droplets, with attaching a droplet or bubble on the cantilever and another on the substrate.

In addition, numerous significant contributions to the theoretical analysis of interaction forces between a solid particle and deformable interface or between two deformable interfaces and the change of deformation during approach have been made. The asymmetric nature of the interaction and the complication of the deformable interface cause mathematical complexity in the interpretation of forces. The interpretation for the deformable interfaces usually involves modeling the drop deformation using the Young Laplace equation [13–15]. Chan et al. [16] developed a sophisticated model using quantities that can be easily obtained from simple experiments and verified by experimental results from AFM.

In this chapter, a direct force measurement of silica nanopaticles between a silica microsphere and an air bubble is performed with AFM. The surface deformability is tuned by the different type and amount of surfactants and the effect of surface deformability on the structuring of colloidal nanoparticles is investigated.

6.2 Results

6.2.1 Force Profiles in the Absence of Additives

The result of a force experiment between a hydrophilic silica microsphere and a bubble in Milli-Q water without extra electrolytes is shown in Fig. 6.1. There, the force is plotted versus relative separation ΔX (change in separation and deformation of the bubble as aforementioned). At ΔX larger than 400 nm, no force was detected and the ΔX was considered as pure separation between the silica probe and the initial bubble surface because soft particles behave as rigid ones when there is no surface force at large distance [17]. A monotonic repulsion began to appear when the probe further approached the bubble. This repulsion is at least partially caused by the electrostatic double layer force because the silica probe is negatively charged and the air-water interface is slightly negatively charged as well [18–20]. The decay length determined in the linear region of the inset logarithmic plot was 102 nm, which agreed with the expected value of the Debye screening length ($\kappa^{-1} = 96$ nm at an ionic strength of 10^{-5} M for pure water).

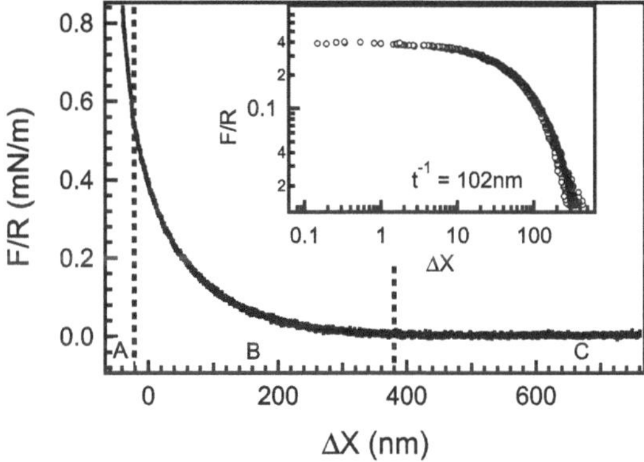

Fig. 6.1 Normalized force (*F/R*) versus ΔX curves of a silica microsphere and an air bubble in water. 'A' presents the constant compliance region where the loading force is linearly increased; 'B' presents the surface force region between the silica microsphere and the *bubble*; 'C' presents the region where no surface force is detected. The monotonic decay region ('B') is fitted with a decay length of 102 nm in the inset graph with double logarithmic scale

When the probe was moved further toward the bubble, the force increased linearly while within the so-called constant compliance region. On solid surfaces, the separation between the silica probe and the substrate does not change in the constant compliance and the increase of force is due to the consistent bending of the cantilever after contacting the solid surface. On the bubble surface it is assumed that the separation between the probe and the bubble surface in the constant compliance region to be not change neither because a stable water film is formed between the silica probe and air [21]. Thus ΔX represents only the deformation of the bubble in the constant compliance region (Eq. 3.7). The deviation of force direction from vertical observed on rigid surfaces is due to the deformation of the bubble from its equilibrium shape. The slope of force versus ΔX at negative ΔX region ($F/\Delta X$) could be used as another measure of the bubble stiffness since $F = k_b \Delta \delta = k_b \Delta X$.

The bubble stiffness of a 800 μm diameter bubble in water calculated from Eqs. 3.9 or 3.10 is typically $k_b = 76\,\mathrm{mN\,m^{-1}}$ which is only two times larger than the spring constant of cantilevers used in the force measurements. Therefore, considering bubble deformation is necessary when measuring forces against bubbles with such soft cantilevers.

6.2.2 Colloidal Nanoparticle Suspensions in the Absence of Surfactants

The normalized force versus ΔX curves for a silica probe interacting with a bubble surface in TMA nanoparticle suspension at varying particle concentrations is shown in Fig. 6.2. When the distance was larger than 200 nm, no force could be detected.

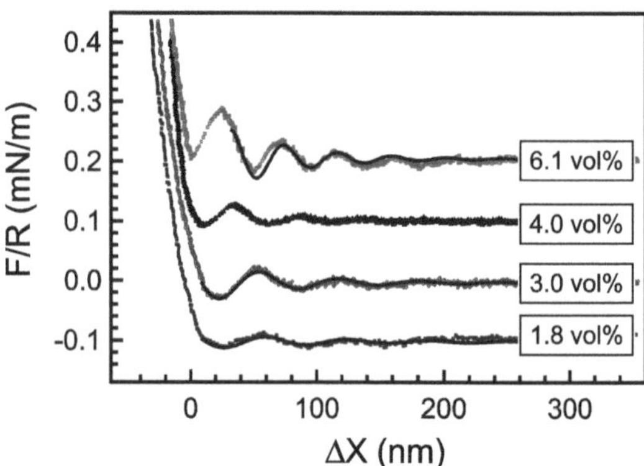

Fig. 6.2 Normalized force (*F/R*) versus ΔX curves of a silica microsphere and an air bubble at different Ludox TMA suspensions (1.8, 3.0, 4.0, 6.1 vol%). The *solid lines* are the corresponding curves fitted to Eq. 2.14. The force profiles have been offset vertically for ease of viewing

The oscillatory force, or structural force of nanoparticles, grew more intense during approach and resulted from the mutual repulsion between the nanoparticles and the layer-by-layer expulsion of the nanoparticles.

The oscillatory wavelengths, which represent the distances between two adjacent nanoparticle layers, decreased with increasing nanoparticle concentrations. This parameter was defined as the distance between successive force maxima or minima. At the same time, the oscillations increased in amplitude at the higher concentrations because the nanoparticles were forced closer to each other, resulting in stronger electrostatic repulsion.

Following the oscillatory force, an attractive depletion force was observed due to the exclusion of all particles from the confined gap between the silica probe and bubble. Additionally, at small separation, an electrostatic repulsive force between confining surfaces was presented, which decayed to zero at larger separation as nanoparticle concentration decreased. This means the phase shift, which can be considered as the depletion zone of the contact layer of particles against the confining surface, increases as particle concentration decreases and exhibit the same behavior as on the solid surfaces.

6.2.3 In the Presence of Non-ionic Surfactants

β-$C_{12}G_2$ is a non-ionic surfactant which adsorbs at the air-water interface resulting in a decrease in the surface tension. The adsorption of β-$C_{12}G_2$ to negatively charged silica has been shown to be weak [22, 23]. The deformation of the bubble at the same nanoparticle concentration with varying β-$C_{12}G_2$ concentration is illustrated in Fig. 6.3a. The slope of the force in the constant compliance region decreased with increasing β-$C_{12}G_2$ concentration. The surface tension of the bubble decreased from 72 to 50 mN m^{-1} at 5×10^{-5} M β-$C_{12}G_2$ and to 40 mN m^{-1} at 10^{-4} M β-$C_{12}G_2$, and the corresponding bubble stiffness was 44 and 35 mN m^{-1}, respectively. The decrease

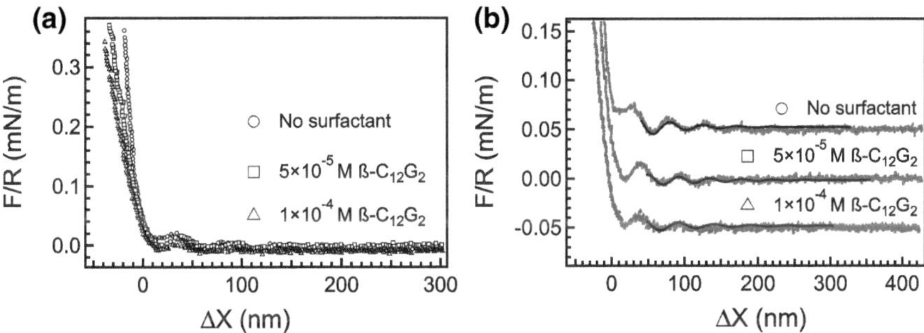

Fig. 6.3 **a** Interaction between a silica microsphere and an air bubble in a 4.9 vol% silica nanoparticle suspension at different β-$C_{12}G_2$ concentrations (0 M, 5×10^{-5} M, 10^{-4} M). **b** The force profiles have been offset vertically for ease in comparison of oscillatory forces. The *solid lines* are the corresponding curves fitted to Eq. 2.14

in the bubble stiffness, or increase in deformability was caused by the decrease of interfacial tension and can be understood by means of Eq. 3.11.

The force profiles as shown in Fig. 6.3b were fitted with Eq. 2.14 in order to obtain the quantitative values of oscillatory wavelength and amplitude. The oscillatory wavelengths showed no change after adding different amount of β-$C_{12}G_2$ surfactant into nanoparticle suspensions. A decrease in oscillatory amplitude with increasing β-$C_{12}G_2$ surfactant concentration was observed due to the reduced surface stiffness and surface charge. The pure air–liquid interface is assumed to be negatively charged [20] and the β-$C_{12}G_2$ molecules partially replace the negative charges. A decrease in surface charge leads to a reduction of the oscillatory amplitude as previously shown in Chap. 5, also in which it is shown that a modification of the charge, or potential of the confining surfaces has no effect on the oscillatory wavelength.

The force profiles of 5×10^{-5} M β-$C_{12}G_2$ at different nanoparticle concentrations are shown in Fig. 6.4. The oscillatory amplitude increased with nanoparticle

Fig. 6.4 Interaction between a silica microsphere and an air bubble at 3.1, 4.6, 6.1 and 9.0 vol% Ludox TMA suspension with 5×10^{-5} M β-$C_{12}G_2$. The *solid lines* are the corresponding curves fitted to Eq. 2.14. The force profiles have been offset vertically for ease of viewing

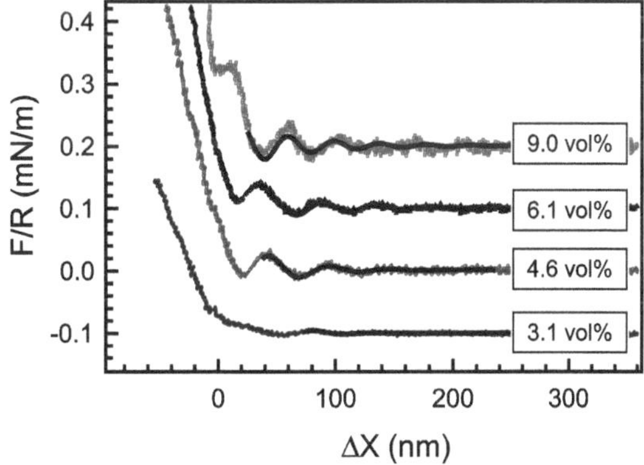

concentration while the oscillatory wavelength decreased since the nanoparticles were closer at the higher concentrations. This behavior was the same as in the absence of added surfactant.

6.2.4 In the Presence of Anionic Surfactants

Sodium dodecyl sulfate is an anionic surfactant which only adsorbs at the air–water interface. A stable film of nanoparticles was formed between the silica probe and the bubble in this case as well, and the repulsive force at the constant compliance region was also observed and attributable to the electrostatic double layer force. The bubble stiffness slightly increased to $80\,\mathrm{mN\,m^{-1}}$ although the interfacial tension did not show measurable change at $5 \times 10^{-5}\,\mathrm{M}$ SDS in comparison to that of pure water (Fig. 6.5a). This can be explained according to Eq. 3.11 which expresses the effect of the decrease of Debye length on the increase of the bubble stiffness. Charged SDS brings extra dissociated ions into the suspension, thus leading to a decrease of the Debye length. An increase of the oscillatory amplitude in the nanoparticle force profile was observed for two reasons. In addition to the slightly increased surface stiffness, it was also likely due to dodecyl sulfate ions adsorbing at the air-water interface and the increase of the interfacial effective charge. Hence, an increase in the electrostatic double layer force with increasing SDS concentration up to the critical micelle concentration (CMC) would be expected [24]. The oscillatory wavelengths obtained after quantitative fitting were found to remain the same compared to the previous cases without surfactant and with β-$C_{12}G_2$ (Fig. 6.5b).

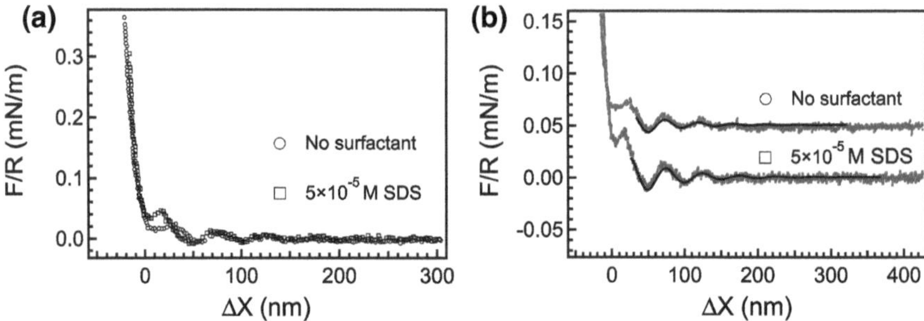

Fig. 6.5 **a** Interaction between a silica microsphere and an air bubble at 4.9 vol% Ludox TMA suspensions with $5 \times 10^{-5}\,\mathrm{M}$ of SDS and without surfactant. **b** The force profiles have been offset vertically for ease of viewing. The *solid lines* are the corresponding curves fitted to Eq. 2.14

6.2.5 *In the Presence of Cationic Surfactants*

Unlike SDS and β-$C_{12}G_2$, hexadecyltrimethylammonium bromide (C_{16}TAB) is a cationic surfactant which not only strongly adsorbs at the air-water interface, but also at the silica microsphere and nanoparticle surfaces, due to interaction of opposite charges on the silica surface and the cationic surfactant head group.

The contact angle measurements of C_{16}TAB on silicon wafer shown in Fig. 6.6 displayed an increase of contact angle to a maximum at around 5×10^{-5} M followed by a decrease again with further increase of C_{16}TAB, which indicated that a monolayer of C_{16}TA$^+$ was formed on silicon wafer at a concentration of 5×10^{-5} M [25]. While at this concentration, the adsorption of C_{16}TAB on the bubble surface was very low and only led to a reduced surface tension of approximately 1.5 %.

Based on the contact angle measurement, the attractive forces due to adsorption of C_{16}TA$^+$ on the silica probe were expected to be measured. A snap into the bubble often occurred in C_{16}TAB solution during manual approach, which increased the difficulty of measurement. An example is shown in Fig. 6.7 when the full piezo range was used in the experiment. A jump-to contact appeared during approach and a big adhesion existed during retraction and no jump-off from the contact was observed. The jump-to contact and the adhesion took place due to the hydrophobic attractive force between C_{16}TA$^+$ adsorbed silica probe and the bubble. However, once nanoparticles were added, the long-range oscillatory force induced a repulsive structural barrier which helped to overwhelm the hydrophobic attraction by forming the layers between the silica probe and the bubble. Thus the AFM force curves in such concentration of C_{16}TAB containing silica nanoparticles could be recorded.

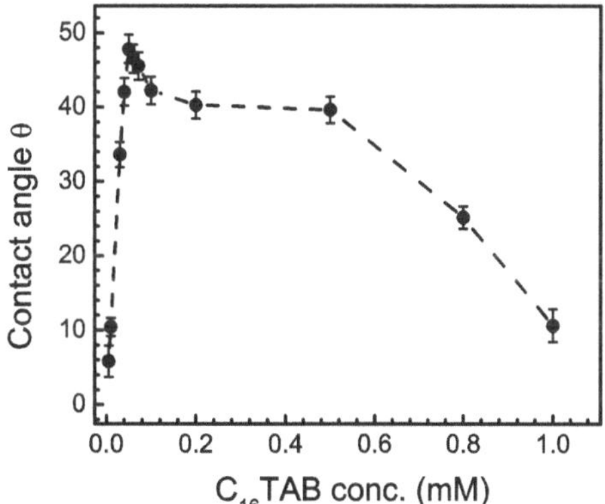

Fig. 6.6 The contact angle of C_{16}TAB on a silicon wafer as a function of surfactant concentration. The maximum of the contact angle appears at a concentration of 5×10^{-5} M resulting from the monolayer formation of cationic surfactant on the negatively charged silica surface. The further decrease of the contact angle is because of the bilayer formation of the surfactant and re-hydrophilization the silica surface

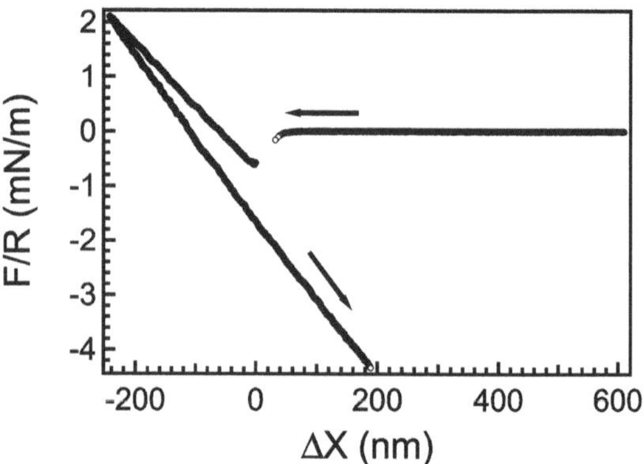

Fig. 6.7 Interaction between a silica microsphere and an air bubble at 5×10^{-5} M of C_{16}TAB solution without colloidal nanoparticles. A jump-in contact appeared during approach and a big adhesion force existed during retraction

Figure 6.8a shows that an oscillatory force began to be detected around 200 nm and was present until a separation of 30 nm. When the last layer of nanoparticles was excluded, the probe was attracted by the bubble due to the hydrophobic force and it penetrated the bubble at a certain short distance. A three-phase contact line (TPC) was formed as a consequence and an increased loading force versus the corresponding deformation of the bubble thus was detected at smaller ΔX. When the probe was retracted, an adhesion force occurred instead of the oscillatory force and resisted the detachment of the probe from the bubble, only the probe was even further retracted at some point, it overcame the adhesion force and was released from the bubble. The amplitude of the adhesion was found to be dependent on the loading force, in general a larger loading force led to a stronger adhesion until the maximum was reached. For adhesion the advancing contact angle was significant, the advancing contact angle could be calculated from

$$\cos \theta_a = \frac{R - D}{R} \tag{6.1}$$

where D is the jumping off distance [8].

The advancing contact angle varied greatly in each force measurement with a maximum of 35° observed, which was smaller than the contact angle measured on the planar silicon wafer in the equilibrium state. This phenomenon probably was due to the unstable organization process of cationic surfactant adsorbed to the silica when surfactant concentration was below the CMC. The slow adsorption of cationic surfactant was reported by Rutland et al. [26] and Fleming et al. [27], who found the build-up of the CTAB layer on a silica surface became more rigid with time.

Figure 6.8b shows the oscillatory forces in the presence and absence of C_{16}TAB. The oscillatory wavelengths in both cases remained constant. This further indicates that the layering distance of nanoparticles in the confinement is particle number density determined, even though the surface charge of the nanoparticles was somewhat reduced after the adsorption of oppositely charged surfactants. The force slope at negative ΔX was lower than in the corresponding case in the absence of surfactant, meaning the deformability of the bubble increased after adding 5×10^{-5} M of

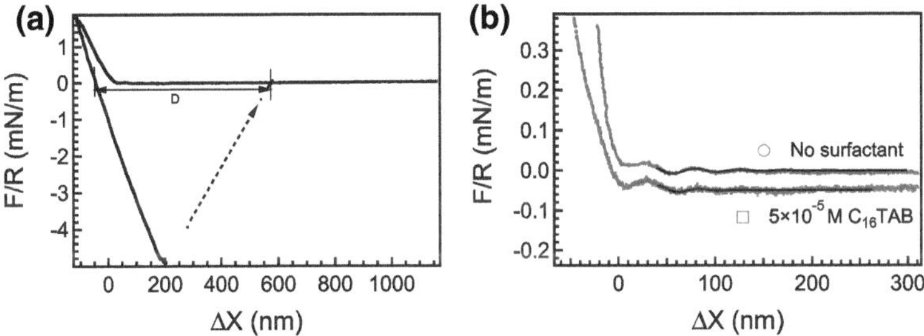

Fig. 6.8 **a** Interaction between a silica microsphere and an air bubble at 4.9 vol% TMA suspensions with 5×10^{-5} M of C_{16}TAB. The oscillatory force of nanoparticles appears during approach while a pronounced adhesion force appears during retraction. 'D' denotes the distance of jumping off contact which is used to calculate the advancing contact angle. **b** The oscillatory forces in the presence and absence of C_{16}TAB are compared. The force profiles have been offset for ease of viewing. The *solid lines* are the corresponding curves fitted to Eq. 2.14

C_{16}TAB, even though the interfacial tension at this concentration was just slightly smaller than that of water. The calculated bubble stiffness was 59 mN m^{-1}, which was attributed to the change in the contact angle as described in Eq. 3.11.

In comparison to the absence of surfactant, a reduction of force amplitude was observed. The reasons were most likely the decreased surface stiffness, and the reduced surface charge both on bubble surface and nanoparticle surfaces.

6.3 Discussion

6.3.1 The Effect of Surface Tension on the Deformability of the Air-Liquid Interface

In contrast to the deformation of elastic and viscoelastic materials, which is controlled by the bulk materials properties, the deformation of bubbles (air–liquid interface) is controlled by the surface tension and the pressure across the interface. The deformation, or elasticity, of the air–liquid interface is typically measured by the oscillating bubble/droplet method. Here, AFM force measurement provides a direct way to determine the deformation of the bubble by assuming that it behaves as a Hookean spring under force applied by AFM probe. Attard and Chan et al. [16, 28] concluded that a Hookean force law is valid for weak forces ($F/2\pi R \ll \gamma$). The existence of the constant compliance region with a linear slope in the force curves is the evidence of linear elasticity for the fluid interface.

The deformability can be expressed as the bubble stiffness by Eq. 3.9, or directly from the slope of force curves in the negative ΔX region with Eq. 3.10. The values of the bubble stiffness, surface tension, oscillatory wavelength, and the oscillatory amplitude at varying Ludox TMA and surfactant concentrations are summarized in

Table 6.1. The increase in the Ludox concentration did not cause significant change in the surface stiffness of the bubble. This was due to the negligible change in the air–liquid interfacial tension with an increase in the particle concentration, although the Debye length of the aqueous solution did decrease. Thus the effect of the Debye length on the surface stiffness of the bubble is assumed to be relatively small.

At a given Ludox concentration (4.9 vol%), the plot of the experimental bubble stiffness versus surface tension is shown in Fig. 6.9a. The square points were obtained from the system with β-$C_{12}G_2$. The increase of β-$C_{12}G_2$ concentration led to the linear decrease of the surface tension. The circle point (SDS) lay along the linear fit because the surface stiffness was not significantly influenced by the decrease of Debye length after adding charged surfactants into this solution. On the other hand, the data for C_{16}TAB (triangle point) deviated from the linear fit because of the increase of the contact angle after the adsorption of C_{16}TAB on the probe surface.

Thus the linear dependency of the bubble stiffness on the air–liquid interfacial tension is valid if the following conditions remain constant: probe radius, bubble radius, Debye length, and the contact angle. This is consistent with the theoretical expression in Eq. 3.11. Also from Eq. 3.11, one should expect that the surface tension and the contact angle play a more important role than bubble size and Debye length, which explains why the decrease in Debye length introduced by the increase of nanoparticle concentration has a negligible effect on the surface stiffness. In addition, the contact angle is strongly associated with the air–liquid interfacial tension, which further supports that the deformation of fluid interfaces is surface tension controlled.

Table 6.1 Summary of the surface tension γ from tensiometer measurements, the bubble stiffness k_b calculated from force curves, the oscillatory wavelength λ, and amplitude A of force curves

Surfactant conc. (M)	ϕ (vol%)	γ (mN m^{-1})	k_b (mN m^{-1})	λ (nm)	A (mN m^{-1})
0	0	71.8	76.7	–	–
	1.8	71.5	75.9	67.8	0.0130
	3.0	71.7	77.0	64.8	0.0373
	4.0	71.9	76.4	54.2	0.0478
	4.9	72.0	75.8	49.6	0.0564
	6.1	72.9	72.7	48.2	0.1018
5×10^{-5} $C_{12}G_2$	3.1	49.7	–	60.7	0.0095
	4.6	49.2	48.1	54.9	0.0253
	4.9	50.2	44.2	50.0	0.0517
	6.1	49.8	45.8	48.2	0.0708
	9.0	50.1	47.3	41.4	0.1115
1×10^{-4} $C_{12}G_2$	4.9	40.1	34.7	50.9	0.0448
5×10^{-5} SDS	4.9	71.5	79.8	50.1	0.0660
5×10^{-5} C_{16}TAB	4.9	70.6	59.3	50.9	0.0500

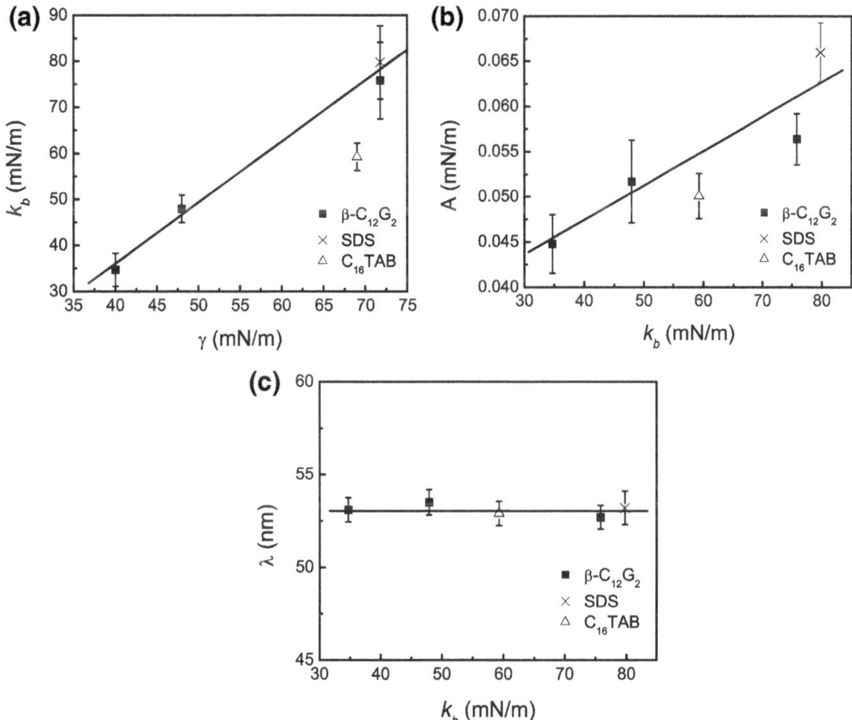

Fig. 6.9 The relationship of the bubble stiffness with the surface tension (**a**) and with the corresponding oscillatory amplitude (**b**) and oscillatory wavelength (**c**) at 4.9 vol% TMA colloidal nanoparticle suspensions

6.3.2 The Effect of Surface Deformability on the Structuring of Nanoparticles

At constant nanoparticle concentration (4.9 vol%), the oscillatory force amplitude exhibited an increase with the bubble surface stiffness (Fig. 6.9b). In the present study, the change of the surface deformability was always associated with the change of the surface charge. Studying the system with non-ionic surfactant β-$C_{12}G_2$, both factors in force amplitude were unable to be analyzed independently, because surface charge and surface stiffness both decreased with increasing surfactant concentration. The reduced surface stiffness and surface charge mutually caused the reduction of force amplitude. Studying anionic surfactant SDS was expected to hopefully shed some light on the two parameters since the increase of surfactant concentration led to an increase of the surface charge. However, a slight increase of the surface stiffness was observed as well, which had the same effect on the change of the force amplitude as surface charge did. Thus the separation of these two causes was also difficult. In the case of cationic surfactant C_{16}TAB, an additional complication was introduced resulting from the interaction between the surfactant and the oppositely charged nanoparticle surface. Therefore, the force amplitude could be considered as the joint consequence of the electrostatic and rigidity effects.

At constant nanoparticle concentration (4.9 vol%), the oscillatory wavelength, representing the layering distance of nanoparticles, showed no dependency on the bubble stiffness or surface deformability (Fig. 6.9c). With regard to the case of cationic surfactant, the oscillatory wavelength did not show any difference even though the surface charge of nanoparticles was reduced additionally. This result is similar to the previous finding on using three different sized nanoparticles, which associate with different surface charge (see Chap. 4). It indicates the electrostatic repulsion can dominate the system over certain surface charge range and thus the particle distance can not be significantly influenced.

The log–log dependence of AFM oscillatory wavelengths versus nanoparticle concentrations is summarized in Fig. 6.10. Wavelengths obtained from measurements of AFM probe against deformable air–liquid interface in the presence and absence of β-$C_{12}G_2$ surfactants were compared to that against a solid silicon wafer. For all deformable cases, the oscillatory wavelength scaled with the nanoparticle concentration as an exponent of -0.33, which agreed very well with the purely space-filling value of $-1/3$. This indicated that nanoparticles under such confinement formed a layered structuring where the interparticle distances scaled to $-1/3$ of the total volume of nanoparticles. These results were in good agreement with the previous experimental results which were based on non-deformable silica surfaces (Chap. 4). The experimental findings indicate that the deformability of the confining surfaces does not change the layered structuring of particles in between. The particle distance remains the same and solely dependents on the particle concentration, or particle number density, regardless of the confinement type. The strength of ordering, however, decreases with increasing surface deformability associated with change of the surface charge, implying less force is needed to exclude particles out of the soft slit-pore.

The difference in wavelengths between the system with the deformable bubble surface and corresponding system with a solid surface was only approximately 1 %. This supports the reliability of using force versus ΔX curves in determining the particle layering distance, even though the deformation of bubble surface contributes at smaller distance.

6.3.3 AFM Versus TFPB

A concentration-independent interparticle distance for charged nanoparticles was observed in TFPB, which contradicted the $\rho^{-1/3}$ scaling law obtained from AFM measurements against an air bubble. The symmetric air–liquid interfaces in the TFPB have larger surface deformability than the one involved in AFM measurements, due to the larger radius of the surface curvature. From AFM measurements the surface deformability shows no significant influence on the characteristic length of

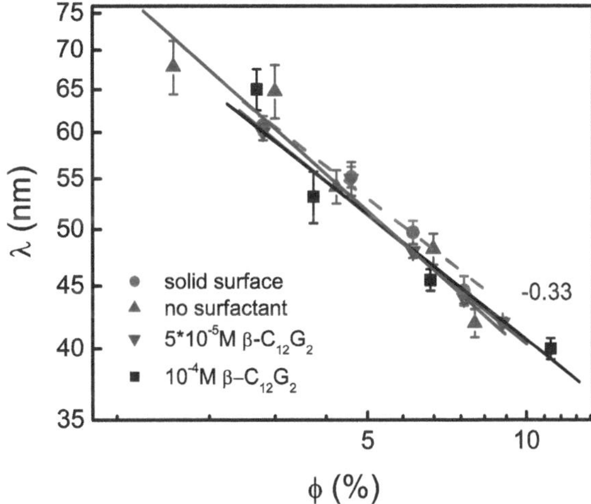

Fig. 6.10 The log–log plot of oscillatory wavelengths versus Ludox TMA concentrations against the air–liquid interface without surfactant, with 5×10^{-5} M of β-$C_{12}G_2$, with 10^{-4} M of β-$C_{12}G_2$, and against a solid silicon wafer

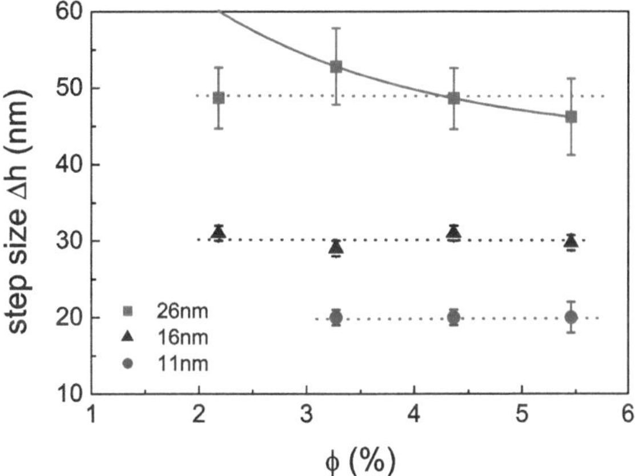

Fig. 6.11 The step size of two adjacent repulsive branch as a function of silica nanoparticle volume fraction, adopted from the previous TFPB work of our laboratory [5]

the structuring. One thus can figure that for symmetric deformable surfaces, the effect of deformability is negligible as well. However, for larger surface deformability the oscillatory forces tilt to negative separation according to the varying deformation extent along the separation distance. In addition, the interparticle distance determined from TFPB is the step size of two adjacent repulsive branches. These two factors cause the inaccuracy in determination. This inaccuracy is dependent on the real interparticle distance, for example, for larger sized particles, the interparticle distance is large and the shift due to the deformation on each peak is therefore smaller in comparison to the particle size than for the smaller ones. This can explain why, for 26 nm sized particles, the interparticle distances can be still somehow described by $-1/3$ scaling law while, for even smaller sized particles, the interparticle distances are totally independent on the concentration (Fig. 6.11).

6.4 Conclusions

AFM provided a direct way to study the structuring of silica nanoparticles confined between a deformable air-water interface and a rigid solid surface. The air–water interface deformability increased with decreasing surface tension and could be observed directly from the change of force slope at the constant compliance region of force profiles.

Three surfactants, β-$C_{12}G_2$, SDS, and $C_{16}TAB$, were used to tune the surface tension thus the surface deformability of air bubble. In the absence and presence of all kinds of surfactants, oscillatory forces of nanoparticles were observed. The only one exception was for cationic surfactant ($C_{16}TAB$), a different behavior was displayed on the retraction part of the force curve, in which a pronounced adhesion appeared. This phenomenon might be attributed to the hydrophobic effect caused by the monolayer formation of cationic surfactant on the silica sphere surface. While with same surfactant, oscillatory force was still observed on the approach branch because the repulsive structural barriers overwhelmed the hydrophobic attraction. Thus a stable thin film of colloidal nanoparticles was assumed to be formed between the silica microsphere and the bubble when strong repulsive interaction existed.

The layering distance and the force strength between nanoparticles could be obtained from the wavelength and the amplitude of the oscillatory force, respectively. It was found that the oscillatory wavelength was not affected by the surface deformability (associated with surface charge) and was the same as between two solid surfaces, while the force amplitude decreased with increasing surface deformability associated with surface charge.

The fact that the surface properties (surface deformability, surface charge) had no effect on the oscillatory wavelength further proved that the layering distance depended solely on the particle concentration. In contrast to this, ordering strength was found to depend on the surface properties as well, thus it was affected not only by nanoparticle concentration, but also by the surface deformability and surface charge after adding extra surfactants.

References

1. Nikolov, A., & Wasan, D. (1989). *Journal of Colloid and Interface Science, 133*, 1–12.
2. Bergeron, V., & Radke, C. (1992). *Langmuir, 8*, 3020–3026.
3. Nikolov, A., & Wasan, D. (1992). *Langmuir, 8*, 2985–2994.
4. Kralchevsky, P., Nikolov, A., Wasan, D., & Ivanov, I. (1990). *Langmuir, 6*, 1180–1189.
5. Mauser, T. Diplomarbeit, TU Berlin.
6. Ducker, W., Xu, Z., & Israelachvili, J. (1994). *Langmuir, 10*, 3279–3289.
7. Butt, H. (1994). *Journal of Colloid and Interface Science, 166*, 109–117.
8. Preuss, M., & Butt, H. (1998). *Langmuir, 14*, 3164–3174.
9. Preuss, M., & Butt, H. (1999). *International Journal of Mineral Processing, 56*, 99–115.
10. Gillies, G., Buscher, K., Preuss, M., Kappl, M., Butt, H., & Graf, K. (2005). *Journal of Physics: Condensed Matter, 17*, S445–S464.

11. Dagastine, R., Stevens, G., Chan, D., & Grieser, F. (2004). *Journal of Colloid and Interface Science, 273,* 339–342.
12. Vakarelski, I. U., Lee, J., Dagastine, R. R., Chan, D. Y. C., Stevens, G. W., & Grieser, F. (2008). *Langmuir, 24,* 603–605.
13. Miklavcic, S., Horn, R., & Bachmann, D. (1995). *Journal of Physical Chemistry, 99,* 16357–16364.
14. Ralston, J., & Dukhin, S. (1999). *Colloids and Surfaces A, 151,* 3–14.
15. Nguyen, A., Nalaskowski, J., & Miller, J. (2003). *Journal of Colloid and Interface Science, 262,* 303–306.
16. Chan, D., Dagastine, R., & White, L. (2001). *Journal of Colloid and Interface Science, 236,* 141–154.
17. Gillies, G., Prestidge, C., & Attard, P. (2001). *Langmuir, 17,* 7955–7956.
18. Collins, G., Motarjemi, M., & Jameson, G. (1978). *Journal of Colloid and Interface Science, 63,* 69–75.
19. Graciaa, A., Creux, P., Lachaise, J., & Salager, J. (2000). *Industrial and Engineering Chemistry Research, 39,* 2677–2681.
20. Ciunel, K., Armelin, M., Findenegg, G., & von Klitzing, R. (2005). *Langmuir, 21,* 4790–4793.
21. Fielden, M., Hayes, R., & Ralston, J. (1996). *Langmuir, 12,* 3721–3727.
22. Zhang, L., Somasundaran, P., & Maltesh, C. (1997). *Journal of Colloid and Interface Science, 191,* 202–208.
23. Lugo, D., Oberdisse, J., Karg, M., Schweins, R., & Findenegg, G. H. (2009). *Soft Matter, 5,* 2928–2936.
24. Schulze, H., & Cichos, C. (1972). *Zeitschrift fuer Physikalische Chemie (Leipzig, Germany), 251,* 252–268.
25. Bijsterbosch, B. (1974). *Journal of Colloid and Interface Science, 47,* 186–198.
26. Rutland, M., & Parker, J. (1994). *Langmuir, 10,* 1110–1121.
27. Fleming, B., Biggs, S., & Wanless, E. (2001). *Journal of Physical Chemistry B, 105,* 9537–9540.
28. Attard, P., & Miklavcic, S. (2001). *Langmuir, 17,* 8217–8223.

Chapter 7
Structuring of Nonionic Surfactant Micelles

7.1 Introduction

Oscillatory forces due to soft colloids, such as surfactant micelles and microemulsion droplets, have been measured by means of thin film pressure balance [1–6], by surface-force apparatus [7, 8], by light-scattering method [9], and by electron cryomicroscopy [10, 11]. Under certain conditions, not the full oscillation, but only the repulsive parts are detectable, which leads to a step-wise thinning or "stratification". These forces can stabilize the liquid films and disperse systems, since they hamper the film drainage [12–16].

Despite the fact that some of the first manifestations of oscillatory forces have been detected with micellar surfactant solutions [17, 18], there are only three applications of CP-AFM to micellar systems [19–21]. Well-pronounced oscillations in the measured force have been detected in two of the studies [19, 20], for micellar solutions of sodium dodecyl sulfate (SDS). In the case of ionic surfactants, such as SDS, the oscillatory forces are essentially affected by the electric double layers around the micelles [17, 18, 22, 23]. The scaling law of $\lambda = \phi^{-1/3}$ has been found to be valid for charged surfactant micelles as well.

The aim of this chapter is to clarify the structuring of uncharged surfactant micelles and the response of the corresponding characteristic quantities with micelle volume fraction, and to test the validity of the scaling law of $\lambda = \rho^{-1/3}$ and $\xi = R + \kappa^{-1}$. Two types of nonionic surfactants with different micellar relaxation time, Brij 35 and Tween 20, are chosen at volume fraction much above CMC. Effect of the surface charge of micelles and deformability (related to the relaxation time) are discussed in detail. The data are analyzed by means of the hard-sphere theoretical model (see Sect. 2.2.2) [24].

Reprinted with permission from: *Oscillatory Structural Forces Due to Nonionic Surfactant Micelles: Data by Colloidal-Probe AFM vs Theory*, Nikolay C. Christov, Krassimir D. Danov, Yan Zeng, Peter A. Kralchevsky, and Regine von Klitzing, *Langmuir*, **2010**, *26*, 915–923. Copyright (2010), American Chemical Society.

Y. Zeng, *Colloidal Dispersions Under Slit-Pore Confinement*, Springer Theses, DOI: 10.1007/978-3-642-34991-1_7, © Springer-Verlag Berlin Heidelberg 2012

7.2 Results and Discussion

7.2.1 Brij 35: Spherical Micelles

Figure 7.1 shows experimental data for 80 mM Brij 35 solution. The speed of approach and retraction was $100\,\mathrm{nm\,s^{-1}}$. The micellar volume fraction $\phi = 0.257$ was taken from the previous study [12]. The theoretical $F(h)/R$ curve in Fig. 7.1 has been drawn without using adjustable parameters by Eqs. 2.29–2.42. The experimental approach and retraction curves for F/R versus h were translated parallel to the horizontal axis until they overlapped with the theoretical curve in the region of greater distances. Such translation is admissible because the experimental zero on the h-axis is determined with a relatively low accuracy. In colloidal probe AFM measurements, the point of contact ($h = 0$) is usually determined as the point at which the linear compliance line reaches zero force. The error is in the order of nanometers and is due to the low spring constant of the used cantilevers. A coupling to an (optical) interferometric method would overcome this problem, but was not used in the present study.

To determine the zero on the axis of distances, the procedure is performed in the following way. The theoretical curves, like those in Figs. 7.1 and 7.2, are independently calculated (no adjustable parameters) at known micelle diameter, d, and volume fraction, ϕ. Next, the experimental data are translated left or right, until the best coincidence with the theoretical curve is achieved. Then, the zero of the theoretical curve is accepted as coordinate origin, $h = 0$, for the experimental data.

If the silica surfaces are covered by surfactant adsorption layers (or dense layers of adsorbed micelles), as observed in the experiments by Ducker et al. [25] the above definition of coordinate origin implies that the surface-to-surface distance, h, corresponds to the separation between the outer ends of the surfactant adsorption layers, rather than between the underlying silica surfaces. Upon further pressing of the two surfaces against each other, it is possible to deform the surfactant layers adsorbed

Fig. 7.1 Normalized force F/R versus distance h for a 80 mM Brij 35 aqueous solution. The points are CP-AFM data; the *arrows* show the direction of measuring motion: approach and retraction. The *solid line* is the theoretical curve. The micelle mean diameter, d, volume fraction, ϕ, and velocity, u, are given in the figure

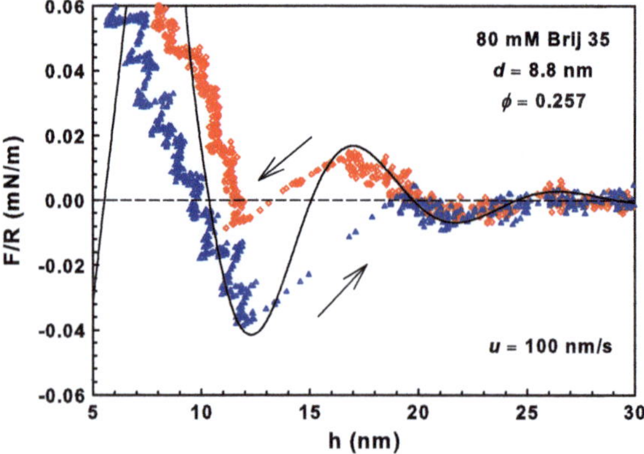

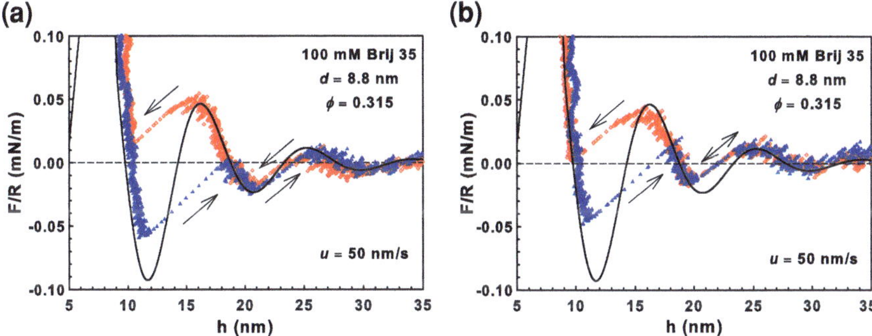

Fig. 7.2 Illustration of the reproducibility of the experimental curves of 100 mM Brij 35 solutions for two different runs (the same cantilever, the same substrate but at two different lateral positions). The points are CP-AFM data for F/R versus h; the *arrows* show the direction of motion; $u = 50\,\text{nm\,s}^{-1}$ is the approach/retraction velocity. The *solid lines* are the theoretical curves.

on the silica. The resulting short-range interaction has been already investigated [25] and it is not a subject of the present study, which is focused on the oscillatory force.

At both approach and retraction, jumps (denoted by arrows in the figures) from one mechanically stable branch of the oscillatory curve to the next one were observed. Such jumps have been observed also in other experimental studies, including foam film studies, where oscillatory forces were detected [1, 12, 19, 20, 26, 27]. Those jumps are due to the relatively low spring constant in comparing to the strong attractive structural force. For the approach curves, the barriers are the oscillatory maxima, whose right branches correspond to mechanically stable states. In contrast, for the retraction curves the barriers are the oscillatory minima, whose left branches correspond to stable states. For the data in Fig. 7.1, the jumps happen close to the tops of the respective barriers. The theoretical and experimental curves are in good agreement except at short distances. At the shorter distances ($h < 12\,\text{nm}$), one micellar layer is trapped between the two solid surfaces and its deformability can be a possible explanation for (i) the difference between the experimental approach and retraction curves (hysteresis) and (ii) some deviations of each of them from the theoretical curve at the smaller h.

To compare the measured oscillatory force with the hydrodynamic interactions, the Taylor formula [28] was used for the force of hydrodynamic interaction between a spherical particle of radius R moving with velocity u toward a planar solid surface

$$F_{Ta} = \frac{6\pi\eta u}{h} R^2 \tag{7.1}$$

as derived in the manner of Eq. 2.8.13 in the literature [29]. As usual, h is the shortest surface-to-surface distance from the particle to the planar solid surface, and η is the viscosity of the liquid phase. The substitution of $\eta = 10^{-3}$ Pa·s, $R = 3.35 \times 10^{-6}$ m, $h = 10\,\text{nm}$, and $u = 100\,\text{nm\,s}^{-1}$ in Eq. 7.1 yields $F_{Ta} = 2.05 \times 10^{-3}$ nN. The value of F_{Ta}/R is equal to one-sixth of the smallest scale division on the ordinate axis in Fig. 7.1. Hence, under the conditions of the present experiments, the hydrodynamic force is negligible in comparison with the magnitude of the oscillatory force.

In Fig. 7.2, the Brij 35 concentration is higher, 100 mM, and the amplitude of the oscillations is larger. The micellar volume fraction, $\phi = 0.315$, was taken from the reported work [12]. Figure 7.2a, b illustrates the reproducibility of the experimental data, which is good except for some differences in the regions of the jumps that have stochastic character. It is interesting to note that in these figures the jumps upon approach happen near the top of the barrier, whereas the jumps upon retraction occur well before the top of the barrier.

In Fig. 7.3, the Brij 35 concentration is 133 mM, the micellar volume fraction was determined from the data fit, which yielded $\phi = 0.401$. The comparison between experimental and theoretical data implies that jumps occur well below the theoretical maxima and above the minima, respectively. The branches have a steeper slope than at lower concentrations (Figs. 7.1 and 7.2), which indicates that the micelle layers are less compressible. At a distance of about 9–10 nm, a strong repulsion was measured, but no further material could be pressed out. Upon retraction, the particle jumps from this first minimum, over the second one, up to the third minimum's stable branch. This behavior can be explained by the fact that a strong attraction between the surfaces leads to a sudden jump-off from the contact; the energy, accumulated during the climbing of the energy barrier, is suddenly released and the system jumps back to a large distance.

Figure 7.4 illustrates the effect of the experimental velocity, u, on the measured force-versus-distance dependence for 150 mM Brij 35 and determined volume fraction $\phi = 0.445$. The latter value is below the Alder phase transition for hard spheres at $\phi = 0.494$, [30, 31] thus the micelles are still considered as spheres.

In Fig. 7.4a, b, the experimental velocities are in the range of optimum velocities for approach and retraction of the colloidal probe against the planar silica surface. Below this range of speeds, the hydrodynamic drift and the noise (that modulates the

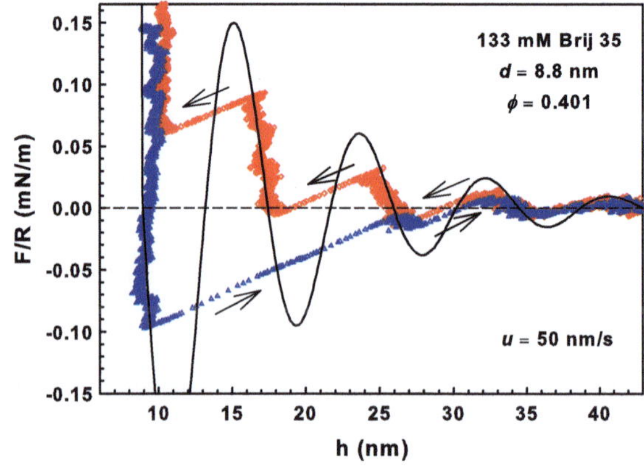

Fig. 7.3 F/R versus h for a 133 mM Brij 35 solution. The points are CP-AFM data; the *arrows* show the direction of motion; $u = 50\,\mathrm{nm\,s^{-1}}$ is the approach/retraction velocity. The *solid line* is the theoretical fit

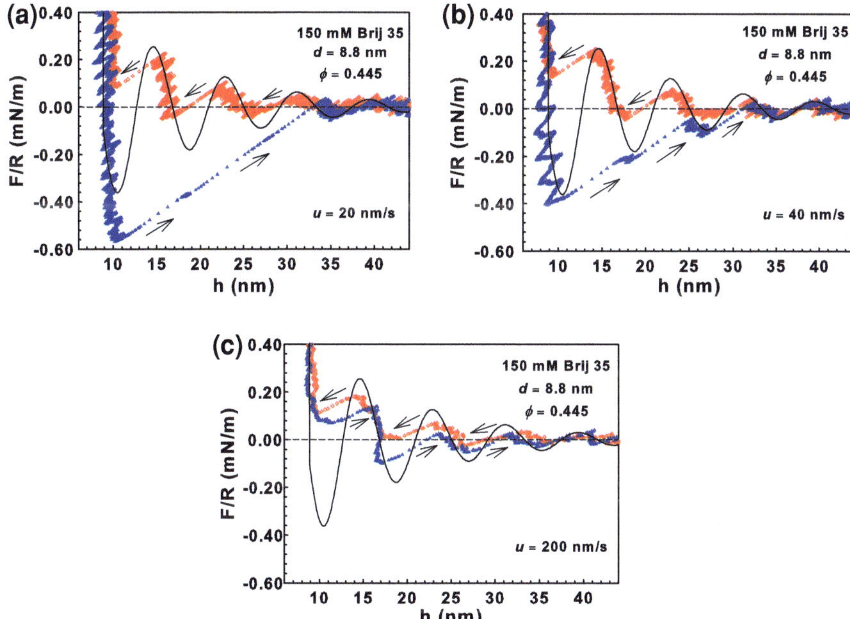

Fig. 7.4 Effect of the rate of measuring motion: force versus distance for 150 mM Brij 35 solutions. The points are CP-AFM data; the *arrows* show the direction of motion. The *solid lines* are theoretical fits. The velocity of the colloidal probe is **a** $u = 20\,\mathrm{nm\,s^{-1}}$; **b** $u = 40\,\mathrm{nm\,s^{-1}}$; **c** $u = 200\,\mathrm{nm\,s^{-1}}$

obtained curves) is too high. Above the optimum speed, the system cannot rearrange fast enough after the expulsion of one layer of micelles. The latter case is illustrated in Fig. 7.4c, where at a greater speed ($u = 200\,\mathrm{nm\,s^{-1}}$) the registered oscillatory amplitude is smaller, which indicates a probably lower degree of structuring of the micelles within the film. In Fig. 7.4a, the first transition upon retraction happens below the theoretical minimum, which indicates adhesion in this special case.

As mentioned above, the comparison of Fig. 7.4c with Fig. 7.4a, b shows that the transitions from one stable-equilibrium branch of the oscillatory curve to the next one happens easier (at smaller magnitude of the applied force) when the velocity u of the colloidal probe is greater. As we known, the oscillatory maxima represent barriers against film thinning upon approach of the colloidal probe, whereas the oscillatory minima represent barriers against film thickening upon retraction. In other words, the system opposes the applied external force tending to minimize the changes produced by it, in agreement with Le Chatelier's principle. In the ideal case of quasi-static probe motion (infinitesimally small u and perfect particle structuring), the transitions should happen at the tops of the respective barriers. However, in the real experiment the colloidal probe moves with a finite velocity u, and the resulting hydrodynamic flow perturbs the micelle structuring. The perturbed structure yields easier, and the transition from one stable branch to the next one occurs at a smaller value of the applied external force, i.e., below the top of the respective quasi-static barrier. This effect is greater at higher speeds of particle motion in agreement with the experimental observations (Fig. 7.4).

Comparing Figs. 7.2, 7.3, and 7.4b show results at more or less the same speed but at different surfactant concentrations, an increase in the slope of the force branches is revealed as micelle concentration increases. This is related to larger amplitude, $w_0 = A/(\kappa T/d^2)$, of the force oscillation at greater micelle volume fraction ϕ (see Table 7.1). As known from previous theoretical studies [32], the wavelength of oscillations, characterized by the dimensionless wavelength $\lambda/d = 2\pi/\omega$, decreases, whereas the decay length (correlation length), $\xi/d = q^{-1}$, increases with the rise of ϕ. To illustrate these effects for the investigated system, in Table 7.1 the values of w_0, λ/d, and ξ/d calculated from Eqs. 2.34–2.36 for the respective ϕ values are listed. One sees that λ/d is close to 1 but still varies in the framework of 16 %. In contrast, the variation of the decay length is much stronger: ξ/d increases with a factor of ca. 3. In other words, the micelle structuring penetrates to distance three times farther from the film surface.

With increasing concentration, the first minimum at a short distance during retraction becomes deeper, which indicates a stronger adhesive depletion force. As a consequence, the systems jumps back to larger distances, as already discussed in relation to Fig. 7.3. This effect becomes stronger at lower speed (compare Fig. 7.4a, c). At a speed of $20\,\mathrm{nm\,s^{-1}}$, the system jumps from the first minimum directly to the fourth minimum, by passing the second and third (Fig. 7.4a).

A general feature of the experimental curves in Figs. 7.1, 7.2, 7.3, 7.4 is that they consist of alternating equilibrium and non-equilibrium portions. In contrast, the theoretical curve represents complete equilibration and it could coincide with the respective experimental curve only at its equilibrium portions. One of the benefits from the comparison of theory with experiment is that it enables one to identify the equilibrium and non-equilibrium portions of the experimental curves. It is clearly seen that the non-equilibrium portions represent jumps from a given branch of the equilibrium theoretical curve to the next one. Because these jumps happen relatively quickly, the experimental curve contains a lower number of points in its non-equilibrium parts, which look thinner in the graphs. This is another way to distinguish between the equilibrium (thicker) and non-equilibrium (thinner) portions of a given experimental curve.

Table 7.1 Micelle diameter d, volume fraction ϕ, the dimensionless oscillatory amplitude w_0, wavelength λ/d, and decay length ξ/d, versus the Brij 35 concentration c_s

c_s (mM)	d (nm)	ϕ (vol %)	w_0	$\lambda/d = 2\pi/\omega$	$\xi/d = q^{-1}$
80	8.8	0.257	1.365	1.070	0.594
100	8.8	0.315	1.700	1.026	0.742
133	8.8	0.401	2.305	0.967	1.059
150	8.8	0.445	2.664	0.938	1.337
200	12	0.483	3.001	0.913	1.759

The experimental curves show a strong repulsion at short distances at about 8–10 nm (Figs. 7.2, 7.3, 7.4). This distance is close to the micelle diameter. The simplest explanation is that a last layer of micelles remains between the two surfaces and cannot be pressed out. At 80 mM Brij 35 (Fig. 7.1), the repulsion is less steep and the distances can be reduced down to 5 nm. This could mean that the micelles in the last layer can be deformed due to the high load and could explain why the hysteresis between approach and retraction is so large. At low concentrations the micelles can be more easily deformed due to more surrounding space, while it is more difficult to deform them in a laterally dense layer of micelles.

7.2.2 Brij 35: Elongated Micelles

The experimental results by Tomsic et al. [33] indicate that at a Brij 35 concentration of 200 mM the micelles are elongated rather than spherical. The dynamic light scattering (DLS) [33] gives a hydrodynamic micelle diameter of $d = 12$ nm. The latter value corresponds to hypothetical spherical micelles that have the same diffusivity as the mean diffusivity of the elongated micelles. The AFM data for concentration of 200 mM Brij 35 are presented in Fig. 7.5. As it could be expected, the experimental curves have oscillatory behavior. The theoretical fit with hard sphere model is possible only at greater distances of the equilibrium portions of the curve, $h > 50$ nm. The obtained value of volume fraction is $\phi = 0.483$ for the best fit.

At shorter distances ($h < 50$ nm), it is impossible to fit the data with Eqs. 2.29–2.42. The wavelength of the theoretical curve for hard spheres is independent of the film thickness. In Fig. 7.5, the theoretical curve, obtained by fitting the data for $h > 50$ nm, is extrapolated at shorter distances and compared with the experimental curves at $h < 50$ nm. This comparison indicates that the measured curves have a varying wavelength, which increases with the decrease of h. In other words, at shorter distances the oscillations are non-harmonic. In particular, the slope

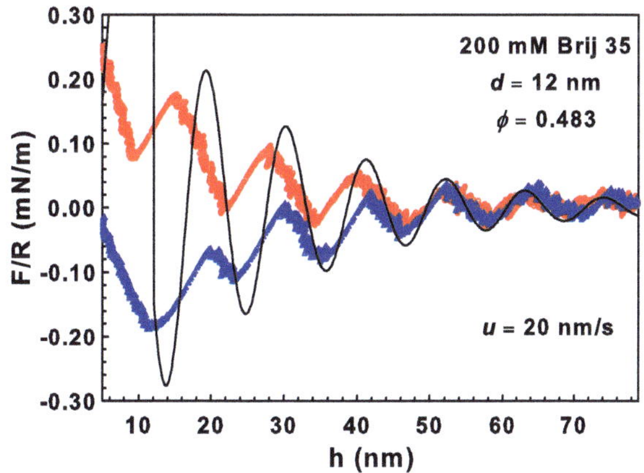

Fig. 7.5 Force versus distance for a 200 mM Brij 35 solution that contains elongated micelles of effective hydrodynamic diameter of $d = 12$ nm. The points are CP-AFM data; the *upper* and *lower experimental curves* are obtained, respectively, at approach and retraction. The *solid line* is the theoretical fit

of the stable branches of the experimental curves is considerably smaller than that of the theoretical curve for hard spheres. Such behavior can be explained with the additional rotational degree of freedom of the elongated micelles. The spatial confinement forces the micelles to orient their long axes parallel to the film surfaces. In such a case, the film thickness can decrease not only by expulsion of micellar layers from the film but also by a gradual reorientation of the elongated micelles. The latter circumstance could explain the observed "softening" of the oscillatory interaction between the two solid surfaces at shorter distances.

7.2.3 Tween 20

The stepwise thinning (stratification) of free foam films from micellar Tween 20 solutions has been investigated [12]. At 200 mM concentration of Tween 20, four steps were registered by the Mysels-Jones porous-plate method [4], and eight steps by the Scheludko-Exerowa capillary cell [34]. Here, the CP-AFM was applied to Tween 20 micellar solutions to directly detect the oscillatory force that engenders the aforementioned stepwise transitions. For Tween 20, the CP-AFM did not detect such well-pronounced oscillatory behavior as with Brij 35 (Fig. 7.1, 7.2, 7.3, 7.4, 7.5). The data in Fig. 7.6 have been obtained at two relatively low force measuring velocities: $u = 5$ and $10\,\mathrm{nm\,s^{-1}}$. As seen in the figure, only one well-pronounced jump has been registered. Among 120 runs, only a few experimental curves were detected that exhibit signs of oscillatory behavior. Below a speed of $5\,\mathrm{nm\,s^{-1}}$, the noise was too large to detect oscillations and above $10\,\mathrm{nm\,s^{-1}}$ also no oscillations were detected. In Fig. 7.6, the experimental curves at approach and retraction (150 mM Tween 20) were compared to the theoretical curve. The values $d = 7.2\,\mathrm{nm}$ and $\phi = 0.250$, determined previously [12], have been used.

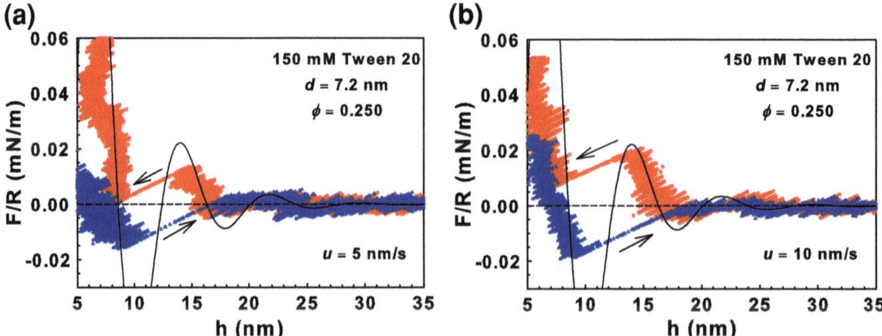

Fig. 7.6 Force versus distance for 150 mM Tween 20 solutions. The points are CP-AFM data for two runs: **a** and **b**; the *arrows* show the direction of measuring motion. The *solid line* is the theoretical curve

Qualitatively similar experimental curves have been obtained for adsorbed micelles [25]. The data in Fig. 7.6 indicate that the experimental force is close to that predicted by the theory for mobile (non-adsorbed) micelles, but the possible presence of adsorbed micelles cannot be ruled out.

The difference between the AFM experimental curves obtained for Brij 35 and Tween 20 micellar solutions indicates that the micelles of Tween 20 are more *labile* and are demolished by the shear stresses engendered by the hydrodynamic flows in the liquid film. Indeed, the lack of oscillatory behavior indicates absence of structural units (i.e., micelles) in the film. The scanning frequencies used in the experiments varied from 0.05 to 0.4 Hz; i.e., the film thinning/thickening continues from 2.5 to 20 s. In contrast, the spontaneous thinning of free films in the capillary cell takes more than 4000 s [12]. In this respect, the capillary-cell method [34] and the thin film pressure balance method [4, 35] are much milder (as compared to the CP-AFM), because the slow hydrodynamic flows in the spontaneously thinning films are accompanied by weak shear stresses that do not cause decomposition of the Tween 20 micelles in view of the well-pronounced stepwise shape of the experimental curves obtained by these methods [12].

The conclusion that the micelles of Tween 20 are more labile as compared to those of Brij 35 is supported by the measured relaxation time of the slow micellar process, τ_2, which characterizes the relaxation of the concentration of micelles in the course of their decomposition to monomers upon a sudden dilution [36, 37]. The stopped-flow dilution technique yields $\tau_2 = 6$ s for Tween 20, and $\tau_2 = 80$ s for Brij 35; see Table 1 in the previous work by Patist et al. [38]. In other words, if the surfactant concentration is suddenly decreased, the perturbations in the concentrations of Tween 20 and Brij 35 micelles exponentially decay with characteristic times of 6 and 80 s, respectively. Hence, the micelles of Tween 20 are destroyed much faster, which is in agreement with the conclusion that they are more labile.

7.2.4 Differences Between Structuring of Nonionic Micelles and Charged Particles

Both types of systems lead to oscillatory force curves. In the case of nonionic micellar solutions, which behave as hard-sphere fluids, the oscillatory wavelength depends relatively weakly on the volume fraction ϕ and is approximately equal to the micelle diameter (see the values of λ/d in Table 7.1). In contrast, for charged particles the wavelength depends much more strongly on particle concentration. In the case of charged silica particles of diameter 11–26 nm, the oscillation wavelength λ shows a strong dependence of the particle volume fraction and scales as $\phi^{-1/3}$. The second difference is the behavior of the decay length. For nonionic micellar solutions, the decay length increases with the micelle concentration. In contrast, for charged particles it decreases with particle concentration and the relation $\xi = R + \kappa^{-1}$ has been found, indicating the decay length is both particle size and ionic strength controlled.

The wavelength and decay length of charged silica particles correspond to the mean particle distance and correlation length in bulk solutions, respectively, obtained from the structure peak of scattering spectra. The experimental results are in good agreement with Monte Carlo simulations using a grand canonical potential and lead to the conclusion that the interactions between the nanoparticles can be described with the simple potential of screened Coulomb interaction (see Chap. 4).

In literature, an effective diameter of a charged particle including the Debye thickness, κ^{-1}, of the counterion atmosphere is defined as the interparticle distance, which means $\lambda = 2(R + \kappa^{-1})$, where R is the particle hydrodynamic radius. The increase of ϕ leads to an increase of the ionic strength due to counterions dissociated from the charged particles followed by a decrease in κ^{-1} and $2(R + \kappa^{-1})$. Note, however, that the above simple expression for λ does not provide quantitative description of the data from experiments and numerical simulations with stratifying films of charged particles. In the salt-free case, the interparticle distance has been found to be smaller than the effective particle diameter, $\lambda < 2(R+\kappa^{-1})$, and does not change significantly with adding extra salts up to 10^{-3} M (see Chap. 4). It indicates that a long-ranged electrostatic repulsion arisen between charged particles due to the overlap of counterion atmosphere at distances even smaller than one Debye length. Even although this repulsion is screened with adding salts, it hampers the approach of two particles as long as it is sufficient. Therefore, different interaction involved in two different systems manipulates the aforementioned opposite behaviors. For non-ionic surfactant micelles, the interaction between micelles is characterized by the hard core potential, micelles behave indeed like hard spheres and λ equals to $2R$. Because of uncharged surfactant micelles used in the measurements, the expression of $2(R + \kappa^{-1})$ approaches $2R$ assuming κ^{-1} equals zero. For charged nanoparticles, the interaction is dominated by long-ranged electrostatic repulsion due to the overlap of counterion atmosphere, thus $\lambda < 2(R + \kappa^{-1})$ has been found in the considered particle concentration range in the salt-free case.

When charged nanoparticle concentration increases to the threshold that electrostatic repulsion is totally screened by the counterions, the hard-sphere behavior is supposed to be observed. This has been proven by the theoretical calculations and simulations [39], thus $\lambda = 2R$ is achieved. At the same time, when nanoparticles behave as hard spheres, decay length does not decrease with particle concentration, strictly speaking, $\xi = R+\kappa^{-1}$ is no longer valid as proposed for low particle concentration regime in the Chap. 4, but rather increase with particle volume fraction [40]. This coincides with the increase of decay lengths of non-ionic surfactant micelles in this Chapter, which behave as hard-sphere fluids. Thus for charged particles the range of the decay length is determined by the range of the electrostatic repulsion in the normal direction, which is controlled by the hard-core repulsion with radius R and the DLVO repulsion with range κ^{-1}. For uncharged ones (totally screened particles or uncharged micelles), an increase in the sample concentration does not change the range of the hard-core repulsion but rather the in-plane ordering of the layers. The larger the concentration, the higher the in-plane ordering, thus the larger the decay length is.

Another difference between micelles and solid particles is in the scan rate during the measurements. While one uses a higher scan rate (several 100s of $nm\,s^{-1}$) to observe oscillations with solid particles [40], the optimum scan rate in the case of nonionic micelles is quite low ($100\,nm\,s^{-1}$ and lower). This is related to the deformability of the micelles. Micelles decompose under larger shear stress engendered by the hydrodynamic flow during fast scan. According to the micelle relaxation time, an optimum scan rate can be defined.

7.3 Conclusions

In the present study, the oscillatory forces in micellar solutions of the nonionic surfactants Brij 35 and Tween 20 were measured using the CP-AFM. These forces cause stepwise thinning (stratification) of foam and emulsion films, and they can stabilize liquid films and disperse systems under certain conditions [12, 13, 15]. Experimental force curves were obtained at both approach and retraction of the colloidal probe. They were compared with the respective theoretical curves that correspond to a hard-sphere model [24].

Spherical micelles were present at low concentration of Brij 35 and harmonic oscillations were observed. The oscillation wavelength is close to the micelle diameter, slightly decreasing with the rise of concentration, while both the amplitude and decay length of the force oscillation increases, indicating an increased in-plane ordering of the micelles (Table 7.1). In addition, the attraction between the surfaces at short distances (the depth of the first minimum) increases with increasing surfactant concentration, which leads to a strong hysteresis between the regimes of approach and retraction. The attraction can be strong enough that several oscillations detected during approach can be jumped over when the cantilever detaches from contact.

The comparison of theory and experiment gives the complete picture of the investigated phenomena and provides new information and understanding of the observed processes. The experiment gives only parts of the stable branches of the oscillatory force-versus-distance dependence, whereas the theoretical model allows us to reconstruct the full curve, which allows a detailed analysis of the micellar ordering. In particular, by superimposing a given experimental curve on the theoretical one, the point of probe/substrate contact (i.e., the zero on the distance axis) could be accurately determined. At $h \approx d$, a strong repulsion is detected which leads to the conclusion that the system cannot overcome the first (the highest) maximum, explained by the fact that one layer of micelles remains between the surface and cannot be pressed out. At low concentration of Brij 35 (80 mM), the surfaces can be approached down to at least 5 nm and the hysteresis is even greater than for higher concentrations (Fig. 7.1). This could mean that at low concentrations the micelles are deformed under the heavy load at short distances.

In the case of elongated micelles, which are present in the Brij 35 solutions at higher concentrations [33], the experimental data do not show a harmonic oscillation anymore (Fig. 7.5). This can be attributed to the circumstance that the film thickness can decrease not only by expulsion of micellar layers from the film but also by a gradual reorientation of the elongated micelles parallel to the film surfaces.

With Tween 20, the experimental curves do not have such well pronounced oscillatory behavior as with Brij 35. This fact indicates that the micelles of Tween 20 are much more labile than those of Brij 35 and are demolished by the shear stresses engendered by the hydrodynamic flows during the thinning or thickening of the liquid film. In contrast, in the case of Brij 35, the micelles are sufficiently stable, and the experimentally-obtained oscillatory curves are in good agreement with the theoretical predictions for a hard-sphere fluid. This behavior correlates with the characteristic times of the slow micellar relaxation process for the two surfactants. In general, an optimum scanning speed is necessary to be defined for the system to rearrange after the expulsion of former layers of the micelles and thus obtain the force profiles.

References

1. Bergeron, V., & Radke, C. (1992). *Langmuir, 8*, 3020–3026.
2. Bergeron, V., Jimenezlaguna, A., & Radke, C. (1992). *Langmuir, 8*, 3027–3032.
3. Exerowa, D., Kolarov, T., & Khristov, K. (1987). *Colloids and Surfaces, 22*, 171–185.
4. Mysels, K., & Jones, M. (1966). *Discussions of the Faraday Society, 42*, 42–50.
5. Marinova, K., Gurkov, T., Dimitrova, T., Alargova, R., & Smith, D. (1998). *Langmuir, 14*, 2011–2019.
6. Bergeron, V., & Radke, C. (1995). *Colloid and Polymer Science, 273*, 165–174.
7. Richetti, P., & Kekicheff, P. (1992). *Physical Review Letters, 68*, 1951–1954.
8. Parker, J., Richetti, P., Kekicheff, P., & Sarman, S. (1992). *Physical Review Letters, 68*, 1955–1958.
9. Krichevsky, O., & Stavans, J. (1995). *Physical Review Letters, 74*, 2752–2755.
10. Denkov, N., Yoshimura, H., Nagayama, K., & Kouyama, T. (1996). *Physical Review Letters, 76*, 2354–2357.
11. Denkov, N., Yoshimura, H., & Nagayama, K. (1996). *Ultramicroscopy, 65*, 147–158.
12. Basheva, E. S., Kralchevsky, P. A., Danov, K. D., Ananthapadmanabhan, K. P., & Lips, A. (2007). *Physical Chemistry Chemical Physics, 9*, 5183–5198.
13. Wasan, D., Nikolov, A., Kralchevsky, P., & Ivanov, I. (1992). *Colloids and Surfaces, 67*, 139–145.
14. von Klitzing, R., & Muller, H. (2002). *Current Opinion in Colloid and Interface Science, 7*, 42–49.
15. Sonin, A., & Langevin, D. (1993). *Europhysics Letters, 22*, 271–277.
16. Wasan, D., & Nikolov, A. (2008). *Current Opinion in Colloid and Interface Science, 13*, 128–133.
17. Nikolov, A., & Wasan, D. (1989). *Journal of Colloid and Interface Science, 133*, 1–12.
18. Nikolov, A., Kralchevsky, P., Ivanov, I., & Wasan, D. (1989). *Journal of Colloid and Interface Science, 133*, 13–22.
19. McNamee, C., Tsujii, Y., Ohshima, H., & Matsumoto, M. (2004). *Langmuir, 20*, 1953–1962.
20. Tulpar, A., Van Tassel, P., & Walz, J. (2006). *Langmuir, 22*, 2876–2883.
21. McNamee, C., Tsujii, Y., & Matsumoto, M. (2004). *Langmuir, 20*, 1791–1798.
22. Blawzdziewicz, J., & Wajnryb, E. (2005). *Europhysics Letters, 71*, 269–275.
23. Klapp, S. H. L., Zeng, Y., Qu, D., & von Klitzing, R. (2008). *Physical Review Letters, 100*, 118303.

24. Trokhymchuk, A., Henderson, D., Nikolov, A., & Wasan, D. (2001). *Langmuir, 17*, 4940–4947.
25. Grant, L., Tiberg, F., & Ducker, W. (1998). *The Journal of Physical Chemistry B, 102*, 4288–4294.
26. Kolaric, B., Jaeger, W., & von Klitzing, R. (2000). *The Journal of Physical Chemistry B, 104*, 5096–5101.
27. von Klitzing, R., Kolaric, B., Jaeger, W., & Brandt, A. (2002). *Physical Chemistry Chemical Physics, 4*, 1907–1914.
28. Horn, R., Vinogradova, O., Mackay, M., & Phan-Thien, N. (2000). *The Journal of Chemical Physics, 112*, 6424–6433.
29. Russel, W., Saville, D., & Schowalter, W. (1989). *Colloidal Dispersions*. Cambridge, U.K.: Cambridge University Press.
30. Balescu, R. (1975). *Equilibrium and Nonequilibrium Statistical Mechanics*. New York: Wiley.
31. Anderson, V., & Lekkerkerker, H. (2002). *Nature, 416*, 811–815.
32. Kralchevsky, P., & Denkov, N. (1995). *Chemical Physics Letters, 240*, 385–392.
33. Tomsic, M., Bester-Rogac, M., Jamnik, A., Kunz, W., Touraud, D., Bergmann, A., et al. (2004). *The Journal of Physical Chemistry B, 108*, 7021–7032.
34. Scheludko, A., & Exerowa, D. (1959). *Kolloid Z, 165*, 148–151.
35. von Klitzing, R. (2005). *Advances in Colloid and Interface Science, 114*, 253–266.
36. Aniansson, E. A. G., Wall, S. N. (1974). *The Journal of Physical Chemistry, 78*, 1024–1030.
37. Danov, K., Kralchevsky, P., Denkov, N., Ananthapadmanabhan, K., & Lips, A. (2006). *Advances in Colloid and Interface Science, 119*, 1–16.
38. Patist, A., Kanicky, J., Shukla, P., & Shah, D. (2002). *Journal of Colloid and Interface Science, 245*, 1–15.
39. Klapp, S. H. L., Qu, D., & von Klitzing, R. (2007). *The Journal of Physical Chemistry B, 111*, 1296–1303.
40. Klapp, S. H. L., Grandner, S., Zeng, Y., & von Klitzing, R. (2008). *Journal of Physics: Condensed Matter, 20*, 494232.

Chapter 8
Conclusion and Outlook

8.1 Conclusion

The interaction between colloids is the key to controlling their stability and structuring. The application of atomic force microscopy on the force measurement gives us the opportunity to investigate the interaction of colloids under one dimensional confinement. The previous AFM work of colloids were mainly focused on studying their structuring between two smooth solid confining surfaces and only on one quantity charactering the structuring: the wavelength of the oscillation. In this thesis three characteristic quantities of the structuring of silica nanoparticles between two smooth solid surfaces were considered and compared with the bulk counterparts, by combining AFM and SAXS two experimental techniques. In the meanwhile, experimental results are compared to those of Monte Carlo simulations [1–3]. AFM measurements on rough surface(s) and deformable surface were further applied to investigate the effect of confining surface properties on the corresponding structuring. In addition, uncharged colloids: non-ionic surfactant micelles were used in AFM to study the effect of surface charge and deformability of the colloids.

Three quantities were extracted from oscillatory force profile of silica nanoparticles between two smooth solid surfaces. These were the wavelength λ, the decay length ξ, and the amplitude A. The first two characteristic lengths were found to correlate well with the mean particle distance $2\pi/q_{max}$ and correlation length $2/\Delta q$, respectively, as obtained from SAXS structural peak. This observation suggests there is no confinement effect on characteristic lengths themselves that represent the structuring, even though the confinement indeed induces a layered structure of the particles. These apparently contrasting results can be understood by considering the in-plane structure and the asymptotic range used for fitting the force curves. At particle concentrations below 10 vol%, no in-plane structure was observed by AFM or Monte Carlo simulations and the fitting based on asymptotic behavior worked well until the first minimum. The fitting did not work for the first maximum, represented as the contact layer, at concentration above 10 vol%. This suggests a possibly higher ordering formed in the contact layer. Nevertheless, the fitting was

Y. Zeng, *Colloidal Dispersions Under Slit-Pore Confinement*, Springer Theses,
DOI: 10.1007/978-3-642-34991-1_8, © Springer-Verlag Berlin Heidelberg 2012

only performed in the asymptotic range, where the particles within the layers were fluid-like, thus the structuring in confinement reflected that of the bulk.

A more quantitative study revealed that the oscillatory wavelength of silica nanoparticles followed the bulk behavior, the relation of $\lambda = \rho^{-1/3}$ was observed irrespective of the particle size (and associated surface charge) and the ionic strength of the solution. The previous description of effective diameter of a charged particle $2(R + \kappa^{-1})$ as the particle distance under confinement was found not quantitatively valid in the present charged system. Instead, the interparticle distance was found to be smaller than the effective particle diameter, $\lambda < 2(R + \kappa^{-1})$. A repulsive interaction is therefore suggested to exist among the silica nanoparticles, and $\lambda = \rho^{-1/3}$ scaling law is a general description for the distance of charged particles in the direction normal to the confining walls, as long as the repulsive interaction is sufficiently long-ranged.

In contrast to the wavelength, which only showed pure volume effect, the decay length was found to be controlled both by the particle size and ionic strength of the solution. A relation of $\xi = R + \kappa^{-1}$ was found at silica particle concentrations below 10 vol%. This relation is supported by the fact that, on one hand, in the low particle concentration regime the range of the correlations is determined by the range of the interaction potential. On the other hand, the range of this potential is determined by the hard-core repulsion with radius R and the DLVO repulsion with range κ^{-1}.

Considering the determination of the Debye length (ionic strength) of the solution, a new method was established to convert the conductivity of the solution into the ionic strength. Instead of using common Russell prefactor, which is valid for simple electrolytes, the individual prefactor for each investigated system can be determined in the linearly dependent regime of conductivity versus ion concentration. The Debye lengths obtained by this new method shows a good agreement with those calculated with Eq. 2.18.

The amplitude A which is a consequence of the strength of interparticle and particle-wall interaction was found to increase linearly with particle concentration. This is attributed to the increased interparticle interaction with narrowing interparticle distance. The amplitude also showed a dependency on the ionic strength of the solution and the particle surface charge. An inverse dependency of amplitude on the ionic strength and a linear dependency on the square of the particle surface charge can be understood by the definition of the electrostatic repulsion: the prefactor $\tilde{Z}^2$ and the interaction range κ^{-1}.

Because there is no direct relation between the strength of interaction obtained from AFM and SAXS, the effect of confinement on the interaction strength was studied between confining surfaces with varied surface potential. An enhanced force amplitude was observed between confining surfaces with higher potential, with wavelength and decay length remaining the same at a given particle concentration. This is explained by the fact that an increase in wall charge strongly changes the screening of the coulomb repulsion between the silica particles and the like-charged confining surface. Monte Carlo simulations [4], based on a modified particle-wall potential with considering the additional wall counterions which accumulate in a thin layer at the wall surface into the particle-wall interaction, yielded a qualitative agreement with the AFM results. For small surface potentials ($0\,\mathrm{mV} \leq |\psi_S| \leq 40\,\mathrm{mV}$) the confining surface potential dominated the particle-wall repulsion and led to an exclusion of particles from the slit-pore. The opposite behavior occurred when $|\psi_S| \geq 40\,\mathrm{mV}$ because the decreased screening length led to an accumulation of particles in the slit-pore. The higher the mean particle density (more particles move from the connected bulk reservoir into the slit), the higher the amplitude is. A decrease in phase shift was accompanied with the increase in amplitude, resulting from the corresponding decrease in the range of particle-wall interactions due to the strongly increased screening of the coulomb repulsion.

A significant reduction in the force amplitude was observed on polyelectrolyte-coated confining surfaces with increasing number of layers and ionic strength of the solutions. Due to a rare change in the corresponding surface potential, this reduction in amplitude was correlated with the change in surface roughness. The surface roughness was found to increase with increasing number of layers as well as the ionic strength, and decreasing the charge density of polyelectrolyte. A phase shift towards a larger separation accompanied the reduction in force oscillation. The roughness-induced reduction in amplitude and shift in phase can be understood as a superposition of oscillatory forces between many surfaces located at slightly different distances. At the roughness threshold, sufficient separation difference among the points on the surfaces smeared out the oscillations and the surface force showed a pure monotonic behavior. A roughness of a few nanometers on a single surface, which corresponded to about $10\,\%$ of the nanoparticle diameter, was sufficient to eliminate the oscillatory force of $26\,\mathrm{nm}$ diameter silica nanoparticles in this study. In order to show an oscillatory force, the particles must be able to be correlated over a reasonably long range. This requires that both the particles and the surfaces have a high degree of order or symmetry. If one of them is missing, so is the oscillation.

Motivated by the lack of dependence of interparticle distance on particle concentration obtained by TFPB, the effect of confining surface deformability on the structuring of silica nanoparticles was studied. An asymmetric confinement was made between a solid silica probe and an air bubble surface and surface deformability was effectively tuned by adding surfactants. The air-water interface deformability increased with decreasing surface tension and could be observed directly from the change of force slope at the constant compliance region of force profiles. Normally, a decreased confining surface charge was associated with the increase in surface deformability. The oscillatory wavelength was found not to be affected by the surface deformability (associated surface charge) and was the same as between two solid surfaces, while the force amplitude decreased with increasing surface deformability, indicating the force required to exclude the layers of particles was less for deformable surfaces. For cationic surfactant ($\mathrm{C_{16}TAB}$), a different behavior was displayed on the retraction part of the force curve, in which a pronounced adhesion appeared. This phenomenon might be attributed to the hydrophobic effect caused by the monolayer formation of cationic surfactant on the silica sphere surface. Thus a stable thin film of colloidal nanoparticles was assumed to be formed between the silica microsphere and the bubble when strong repulsive interaction existed. The fact that the surface properties (surface deformability, surface charge) had no effect on the oscillatory wavelength further proved that the layering distance depended solely on the particle concentration. In contrast to this, ordering strength was found to depend on the surface properties as well, thus it was affected not only by nanoparticle concentration, but also by the surface deformability and surface charge after adding extra surfactants.

The unchanged wavelength and decay length between confining surfaces of various conditions at a given particle concentration confirm that the characteristic lengths which represent the particle structuring are particle-particle interaction dependent. In contrast to this, surface property-dependent ordering strength are both interparticle and particle-wall interaction controlled.

The scaling law of $\lambda = \rho^{-1/3}$ for oscillatory wavelength and the relation of $\xi = R + \kappa^{-1}$ for correlation length break down in the case of non-ionic surfactant micelles. In the case of spherical micelles, with increasing surfactant concentration, both the amplitude and decay length increased which indicated increasing in-plane ordering of the micelles, while the wavelength of micelles remained the same as the value of micelle diameter. This is because the interaction is characterized by the hard core of the micelles. Thus the wavelength is the diameter of micelles and not affected by the bulk concentration and the ordering is enhanced by pressing more micelles into the layers with remaining the number of layers constant at a given wall separation. Other difference from the oscillatory force curves of charged silica nanoparticles was that a strong hysteresis between the regimes of approach and retraction was observed due to the attraction between the surfaces at short distances. By superimposing a given experimental curve on the theoretical one based on the hard sphere potential [5], the point of probe/substrate contact could be accurately determined. At a separation close to the micelle diameter, a strong repulsion was detected which led to the conclusion that the system could not overcome the first maximum and one layer of micelles remained between the surfaces and could not be pressed out. In the case of elongated micelles, the experimental data did not show a harmonic oscillation anymore. This can be attributed to the circumstance that the film thickness can decrease not only by expulsion of micellar layers from the film but also by a gradual reorientation of the elongated micelles parallel to the film surfaces. In addition, the relaxation time of micelles plays an important role in displaying the oscillatory forces as well. Micelles with short relaxation time do not have well pronounced oscillatory behavior. This can be understood as they are much more labile and are demolished by the shear stresses engendered by the hydrodynamic flows during the approach and retraction. Therefore, an optimum scanning speed is necessary to be defined for the system to rearrange after the expulsion of former layers of the micelles and thus obtain the force profiles.

8.2 Outlook

During the preparation of this thesis, some new questions have arisen that could be further investigated. Due to low concentrations of charged particles examined in this work and large separation range for fitting the force curves, the characteristic lengths of the oscillation correlate with those in bulk, which is isotropic and fluid-like, although particles form layers in the vicinity of the confining surfaces. Deviation from the asymptotic behavior was observed at silica particle concentration higher than 10 vol%. This suggests a higher in-plane ordering. A possible direction for further research is to induce further in-plane structuring. This can be done by increasing the concentration of dye-doped silica particles and pressing them to higher densities to obtain defined in-plane structure, i.e. hexagonal or cubic structure. In the meantime, fluorescence microscopy can be used to follow the particles' dynamics and determine the corresponding structure. The phenomenon of nanoparticle structure formation under confinement is of considerable interest in both science and technology. The nature of the oscillatory structural forces should be further explored to learn how to optimize the interaction patterns of nanoparticles in order to engineer nanomaterials and devices such as quantum dots and quantum wires.

Another open question is the difference between results from TFPB and AFM air bubble measurements. A concentration-independent interparticle distance for charged nanoparticles was observed in TFPB, which contradicted the $\rho^{-1/3}$ scaling law obtained from AFM. Although the uncertainties in determining the step sizes due to the contribution of surface deformability can be considered as one reason, one could also relate this issue with the different packing of particles at the air-liquid interface. During the air bubble measurement, no packing of silica nanoparticles at the interface was found due to the high hydrophility of the particle surface. Thus a possible research direction is that one can tune the surface hydrophobicity to induce the packing. The symmetry of the confinement might be also play an important role in the packing, thus the attachment of a bubble on the cantilever is necessary. The use of additional surfactants for stabilizing the packing needs to be considered as well. An understanding of interface self-assembly and the interaction between particles and surfactants can be explored for a variety of applications including drug delivery.

References

1. Klapp, S. H. L., Grandner, S., Zeng, Y., & von Klitzing, R. (2008). *Journal of Physics: Condensed Matter, 20*, 494232.
2. Klapp, S. H. L., Zeng, Y., Qu, D., & von Klitzing, R. (2008). *Physical Review Letters, 100*, 118303.
3. Klapp, S. H. L., Grandner, S., Zeng, Y., & von Klitzing, R. (2010). *Soft Matter, 6*, 2330–2336.
4. Grandner, S., Zeng, Y., von Klitzing, R., & Klapp, S. H. L. (2009). *Journal of Chemical Physics, 131*, 154702.
5. Christov, N. C., Danov, K. D., Zeng, Y., Kralchevsky, P. A., & von Klitzing, R. (2010). *Langmuir, 26*, 915–923.

Curriculum Vitae

Personal

Name:	Yan Zeng
Place of Birth:	Nanchang (China)
Date of Birth:	August 11th, 1983
Nationality:	Chinese
Address:	1050 Engineering Building I, NC State University, Raleigh, USA
Contact:	yzeng6@ncsu.edu

Education

Technische Universität Berlin, November 2011
Ph.D., *magna cum laude*, Physical Chemistry
Advisor: Prof. Dr. Regine von Klitzing

Humboldt Universität zu Berlin, October 2006
M.S., *very good*, Polymer Science

Wuhan University, June 2004
B.S., Applied Chemistry

Research and Teaching Experience

North Carolina State University, 2012-Present
Post-doctoral research with Prof. Orlin Velev
Topic: Microfibrilated cellulose based materials and composites

Y. Zeng, *Colloidal Dispersions Under Slit-Pore Confinement*, Springer Theses, 119
DOI: 10.1007/978-3-642-34991-1, © Springer-Verlag Berlin Heidelberg 2012

Technische Universität Berlin, 2006–2011
Graduate Student with Prof. Dr. Regine von Klitzing
Ph.D. Thesis: Structuring of Colloidal Dispersions in Slit-pore Confinement

Technische Universität Berlin, 2005–2006
Research Assistant with Prof. Dr. Gerhard Findenegg
Master's Thesis: DSC Study of Melting and Freezing of Organic Substances in
Ordered Mesoporous Silica Materials

Wuhan University, 2004
Research Assistant with Prof. Naicai Cai
Bachelor's Thesis: Chemical Deposition of Polyaniline on Activated Carbon for
Electrochemical Capacitors

Technische Universität Berlin, Spring 2007/2008/2009
Teaching Assistant
Introductory to the experimental part of master's course "Polymers at interface"

Awards

Section Condensed Matter (SKM) Dissertation Award of DPG 2012

Scholarship for female Ph.D. student (Frauen-Promotionsabschlussstipendium),
Technical University Berlin, 2010

Excellent undergraduate's thesis, Wuhan University, 06/2004

Outstanding undergraduate student scholarship, Wuhan University, 2001-2003

Skills

Chemistry: Mesoporous materials synthesis, layer-by-layer deposition, surface
modification, polymerization, self-assembly of colloidal particles

Expert instrumental kills: AFM, including surface imaging, force measurements,
and mechanical properties determination; SAXS; DSC; QCM

Other characterization techniques: Nitrogen adsorption, surface tension, zeta
potential, contact angle, ellipsometry, IR, UV-Vis spectroscopy, cyclic
voltammetry, X-ray diffraction, SEM, TEM

Software: Igor Pro, Latex, MS office, Origin

Language: English (fluent), German (basic), Chinese (native)

Selected Conference Contributions

76th DPG Spring Meeting of the Condensed Matter Section, March 25th–30th 2011, Berlin

25th Meeting of European Colloid and Interface Society September 4th–9th 2011, Berlin

7th Liquid Matter, June 27th–July 1st 2008, Lund, Sweden

72th DPG Spring Meeting of the Condensed Matter Section, February 25th–29th 2008, Berlin

References

Prof. Dr. Regine von Klitzing
Department of Chemistry
Technical University Berlin, Germany
+49-30-314 23476
Email: klitzing@mailbox.tu-berlin.de

Prof. Dr. Gerhard Findenegg
Department of Chemistry
Technical University Berlin, Germany
+49-30-314 24171
Email: findenegg@chem.tu-berlin.de

Prof. Dr. Sabine Klapp
Department of Theoretic Physic
Technical University Berlin, Germany
+49-30-314 23763
Email: klapp@physik.tu-berlin.de

MIX
Papier aus verantwortungsvollen Quellen
Paper from responsible sources
FSC® C105338

If you have any concerns about our products,
you can contact us on
ProductSafety@springernature.com

In case Publisher is established outside the EU,
the EU authorized representative is:
Springer Nature Customer Service Center GmbH
Europaplatz 3, 69115 Heidelberg, Germany

Printed by Libri Plureos GmbH
in Hamburg, Germany